AF864035

Environmental Activism in China

Major environmental degradation is a serious problem for China as the country's economy continues to grow at a phenomenal pace. In recent years, environmental organizations have begun to emerge in China, and in some cases have had remarkable success in affecting policies that would have had significant adverse impacts on the environment. This book, based on extensive original research, adopts a multi-disciplinary research approach to examine environmental activism in China, focusing on four cities. It analyses the nature, characteristics, strategies, organizational modes, and influence of what could be labeled a Chinese environmental movement in the making. In particular, this volume highlights the specificities of Chinese environmental activism in an increasingly globalizing world, along with a comparison with the environmental movement in Western Europe and North America.

Lei Xie received her PhD from Wageningen University, the Netherlands. She is a Honorary Visiting Fellow at the Department of International Politics, City University, London. She has worked in the NGO Research Center at Tsinghua University in Beijing as a research officer.

China Policy Series
Series Editor: Zheng Yongnian, *China Policy Institute, University of Nottingham, UK*

1 **China and the New International Order**
Edited by Wang Gungwu and Zheng Yongnian

2 **China's Opening Society: The non-state sector and governance**
Edited by Zheng Yongnian and Joseph Fewsmith

3 **Zhao Ziyang and China's Political Future**
Edited by Guoguang Wu and Helen Lansdowne

4 **Hainan – State, Society, and Business in a Chinese Province**
Kjeld Erik Brodsgaard

5 **Non-Governmental Organizations in China: The rise of dependent autonomy**
Yiyi Lu

6 **Power and Sustainability of the Chinese State**
Edited by Keun Lee, Joon-Han Kim and Wing Thye Woo

7 **China's Information and Communications Technology Revolution: Social changes and state responses**
Edited by Xiaoling Zhang and Yongnian Zheng

8 **Socialist China, Capitalist China: Social tension and political adaptation under economic globalisation**
Edited by Guoguang Wu and Helen Lansdowne

9 **Environmental Activism in China**
Lei Xie

Environmental Activism in China

Lei Xie

Routledge
Taylor & Francis Group
LONDON AND NEW YORK

First published 2009
by Routledge
2 Park Square, Milton Park, Abingdon, Oxon OX14 4RN

Simultaneously published in the USA and Canada
by Routledge
270 Madison Ave, New York, NY 10016

Routledge is an imprint of the Taylor & Francis Group, an Informa business

© 2009 Lei Xie

Typeset in Times New Roman
by Taylor & Francis Books
Printed and bound in Great Britain
by The MPG Books Group

All rights reserved. No part of this book may be reprinted or reproduced or utilised in any form or by any electronic, mechanical, or other means, now known or hereafter invented, including photocopying and recording, or in any information storage or retrieval system, without permission in writing from the publishers.

British Library Cataloguing in Publication Data
A catalogue record for this book is available from the British Library

Library of Congress Cataloging in Publication Data
Xei, Lei.
Environmental activism in China / Lei Xei.
p. cm. – (China policy series ; 9)
Includes bibliographical references and index.
1. Environmental policy–China. 2. Environmentalism–China. 3. China–Environmental conditions. I. Title.
HC430.E5X45 2009
333.720951–dc22

2008044649

ISBN 978-0-415-47869-4 (hbk)
ISBN 978-0-203-87913-9 (ebk)

For Xavier

Contents

Figures

Maps

Tables

Acknowledgements

This book is based on my doctoral thesis, a fantastic journey studying the Chinese environmental movement. Five years ago, when being introduced to a self-organized voluntary group based on residential communities, I was amazed by their drive in dealing with environmental issues and became very interested in their organization of collective activities. In the ten months during which I conducted field work in China (2003–04), I witnessed not only true passion, but also commitment toward the environment from Chinese individuals, which forms a potential force in China's environmental protection. As amazing as this study is, it also enriched my personal experience. In my field work I encountered and collected not only meaningful data, but also ideas and ideals that still enlighten me now. I have to thank those who provided strong support during my field work and with whom I became friends: Mr Wen Bo, Mr Fu Tao, Mr Gong Yi, Mr Liu Detian, Ms Yun Jianli, Mr Li Zhihe and Mr Sun Jun.

I owe many thanks to Professor Arthur P. J. Mol, with whom I was very lucky to work. Thanks to him, I was greatly enlightened and learned about what a social scientist is. Professor Wang, Ming is another person to whom I owe gratitude. He guided me in recognizing, studying and providing support to Chinese NGOs. I have been very grateful to Dr Hein-Anton Van der Heijden and Professor David Sonnenfeld, who advised me on my research with no limits. I am also thankful to colleagues in the Environmental Policy Group, Wageningen University, with whom I discussed my research in great length, among them are Gert Spaargaren, Zhang, Lei, Astrid Hendriksen, Loes Maas and Michel De Krom.

Abbreviations

ACEF	All-China Environment Federation
ADB	Agriculture and Development Bureau
ASGP	Association of Saunders's Gull Protection
BBEC	Beijing Brooks Education Center
BCA	Bureau of Civil Affairs
BNCEC	Beijing New Citizen Education Center
CCP	Chinese Communist Party
CCTV	Chinese Central Television
CEEC	Center for Environmental Education and Communications
CI	Conservation International
CPC	Chinese People's Congress
CPPCC	Chinese People's Political Consultative Conference
EEERC	Earthview Environment Education and Research Center
EIA	Environmental Impact Assessment
ENGO	environmental NGO
EPB	Environmental Protection Bureau
EPECCC	Environmental Protection and Ecological Construction Coordinating Committee
FON	Friends of Nature
GC	Grassroots Community
GCR	Green Cross
GEI	Global Environmental Institute
GEV	Green Environmental Volunteers
GRW	Green Web
GVB	Global Village Beijing
GDP	gross domestic product
GH	Green Han River
GONGO	government-organized NGO
IED	Institute of Environment and Development
INGO	international NGO
IRN	International River Net
MCA	Ministry of Civil Affairs
MEP	Ministry of Environmental Protection

NDRC	National Development and Reform Commission
NGO	non-governmental organization
NRPB	Natural Resource Protection Bureau
OECD	Organization for Economic Co-operation and Development
OIA	Office of Implementing Agenda 21
OPA	Office of Petroleum Affairs
POS	Political Opportunity Structure
SCCO	Spiritual and Civilization Construction Office
SBCO	Shanghai Bureau of Civil Organizations
SEPA	State Environmental Protection Agency
SEPB	Shanghai Environmental Protection Bureau
SGSF	Shanghai Green Student Forum
SMO	Social Movement Organizations
SMPC	Shanghai Municipal People's Congress
SMS	Social Movement Sector
SMYL	Shanghai Municipal Youth League
SNJI	Shanghai New Jiangwan City Investment Enterprise
SSCPA	Shinestone Community Participation Action
STHWRP	Shuangtai Hekou Wetland Reservation Park
SWAPA	Shanghai Wild Animal Protection Association
SWBPA	Shanghai Wild Birds Protection Association
XCBCA	Xiangfan City Bureau of Civil Affairs
XCPPCC	Xiangfan City People's Political Consultative Conference
XCPC	Xiangfan City People's Congress
WCS	Wildlife Conservation Society
WWF	World Wildlife Fund
UGC	University Green Camp
UNESCO	United Nations Educational, Scientific and Cultural Organization
UNDP	United Nations Development Programme
YMCA	Young Men's Christian Association
ZBEPB	Zhabei District Government Environmental Protection Agency
ZBYL	Youth League in Zhabei District

Introduction

Emerging environmental activism

Since at least 1949, the Chinese state has been rather constrained toward the organization of citizens in non-governmental organizations (NGOs) that have no strong relation to the state or the Communist Party. While in the late 1960s and early 1970s most of the more industrialized countries experienced the emergence of a strong civil society sector focused on environmental issues, and in many of the industrialized and developing countries similar environmental civil society sectors were formed in the 1980s, China lagged behind. It is only over the past decade, since the mid-1990s, that a blossoming of environmental NGOs (ENGOs) can be witnessed in China. By participating in these ENGOs and in various citizens' groups, the public begins to play a role in Chinese environmental governance at various levels. Some examples can shed light on the diversity and development of ENGO activities that are evolving at present in China.

Since the late 1990s, several large-scale environmental campaigns have been led by ENGOs, among which are Beijing's environmentalists. In the construction of riverbanks in 2000, Beijing's Bureau of Water Resources planned to reconstruct the riverbanks of the Kunyu River. The idea was to cement the riverbanks in the northern part at Kunyu in Beijing, which would greatly endanger the abundant natural ecology along the riverbanks. The decision and plans met with strong protests by environmentalists, scholars and local inhabitants. Through the efforts of ENGOs, newspapers, and radio broadcasting, public opinion on this issue was raised. A public hearing was organized by environmentalists to meet top officials of concerned agencies, including the head of the Bureau of Water Resources and the Vice-Mayor in charge of this matter. In this dialogue, the public, led for the first time by environmentalists, articulated their opposing opinion directly to policy-makers. However, the Mayor repressed the outcome of this public meeting by banning any media reporting on this activity. Subsequently, the construction of the riverbanks was implemented as planned. Nevertheless, the environmentalists demonstrated their potential and impact in participating in environmental policy-making.

Early in 2003, another environmental campaign protested against a hydroelectric power project, which was planned in Yang Liuhu in Sichuan Province. Initiated by the Du Jiangyan Management Bureau, the Yang Liuhu dam project of around 1200 meters high was 1310 meters away from the heart of Du Jiangyan—a construction ranked as World Heritage.[1] Cooperation was established among environmental activists, the mass media, and staff who work in the local World Heritage Protection Office of Du Jiangyan. Furthermore, the United Nations Educational, Scientific and Cultural Organization (UNESCO) was contacted through its staff in Beijing, and exerted pressure on the central government. After five months of ENGO activities and protests, in which various publications appeared in Chinese newspapers, this project was halted following a collective decision issued by the Construction Bureau of the central government and the provincial government of Sichuan.

The anti-Yang Liuhu dam campaign witnessed the starting point of similar movements protesting against hydro-power construction projects in southwestern provinces. Among these, the successful protest against the Nu River dam construction is a very influential example in recent years. The original project—planned by the National Development and Reform Commission—appeared as the largest hydro-power project of the past two decades, even larger than the now famous Three Gorges Dam. The campaign against the Nu River project is characterized by a series of collective actions by ENGOs, such as consensus-building through a great number of newspaper publications and TV broadcasts, the building of websites across the country, the mobilization of international pressure on the Chinese government, and cooperation with environmental authorities.

An emergence of environmental activism can be seen not only in many major cities across the country, but also in the formation of collective actions at the national level. In 2005, an open letter was signed by 56 ENGOs to support the Ministry of Environmental Protection (MEP)'s[2] sanction of 30 large infrastructure projects across the country, due to their not having a proper environmental impact assessment.[3] This event was organized and concluded in three days, with the news being reported in a large number of national newspapers and on the internet, including foreign websites, and also reported abroad. This action raised the public's environmental awareness, produced impact on environmental policies, and had significant social leverage. It also pointed out the growing possibilities and room for maneuver for ENGOs to be engaged in Chinese environmental politics.

Despite the growing significance of environmental campaigns and activities conducted by ENGOs, the Chinese environmental movement is still relatively weak compared with what can be witnessed in other countries of more or less similar size and development level (such as India and Brazil). Their weakness is usually illustrated by several characteristics: the rather small number of ENGOs, the limited resources and capacities that these NGOs have, their fairly narrow spectrum of strategies and approaches to combat environmental pollution and natural resource extraction, and their rather informal and undifferentiated internal organizational structure. In Chapters 1 and 2, several of these characteristics are further explored and quantified. In general, three explanations are given for these characteristics of the contemporary environmental movement in China: the short history of the environmental movement, the political context of the authoritarian Chinese state, and the specifics of the Chinese cultural context. I elaborate briefly on the first reason below, to be followed by more extensive elaborations on the second and third reasons in the two following sections.

Young social organizations of any kind at any location, including ENGOs in China, are limited in their capacities and resources, and often show low levels of formality and differentiation in their organizational structure. In China, we see that many ENGOs lack financial and other resources, office space, supplies, and skilled personnel (Wang, M., 2000), which strongly affects their organizational development (Turner, 2003). A large number of them have low levels of formality and organizational structure: they have barely established rules, procedures, and regulations to manage their organizations internally. If the recent history of ENGOs in China is the only or main reason for their weaknesses, one can expect that in a number of years many of these organizations will become professional, established, and mature organizations, with significant resources and capacity. But with a further two potential explanations, the future of the Chinese environmental movement becomes less predictable.

Political constraints on movement activism

Under the rule of the Chinese authoritarian government, environmentalists have been constrained in various ways in their activities, development, and organizational forms.

The development of the Chinese environmental movement and green activism is closely related to, and structured by, the specific political environment in China. In general, the Chinese authorities are not keen on encouraging organized efforts from the public, which are often associated with political dissent. The state has established restrictive administrative procedures to govern NGOs. They are particularly strict in the registration of voluntary organizations. A dual supervising mechanism is utilized in order to control the growth of voluntary groups and organizations, which requires each NGO to find an organization as its mother-in-law institution. At the same time, these civil society organizations remain under the regulation of the Bureau of Civil Affairs. For instance, by requiring high amounts of initial funds, the political authorities limit the registration of NGO as social organizations.[4] The political authorities also regulate the areas of activities in which voluntary or non-governmental organizations may work.

Besides restrictive regulations on the establishment and development of civil organizations, the state plays a monopoly role in Chinese ecological governance, which provides limited political channels for public participation. At present, relatively few institutional arrangements exist for ENGOs to participate in policy-making and implementation on the environment. The hostile attitude of the political authorities toward the organization of voluntary organizations and the limited institutional opportunities for political participation constrain public involvement in environmental governance. It is only very recently that the Chinese government has begun to encourage the public's involvement in ecological governance. For instance, in the Environmental Impact Assessment Law (2003), consultation with the public is required for initiation of environmental policies and projects. Yet, lacking specific prescription, it is only partially implemented. Constraints also exist on public access to information. Chinese authorities try to control information flows on major environmental issues, especially those of significant political sensibility. Environmental information and news releases to the public lack transparency and interdependency, although the situation is improving and is already significantly better than on issues such as the Falun Gong, Taiwan, or national security (Mol, 2006; Mol & Carter, 2006).

Chinese environmentalists have to have relatively mild attitudes toward their government and governmental authorities, and refrain from too openly debating, criticizing, or confronting the state's environmental policies (Saich, 2000), in order to avoid being labeled as 'political' activists. Consequently, they can rarely criticize strategic decision-making, issues related to human rights, or major economic development strategies. This also restricts the action repertoire of ENGOs; where highly visible confrontational activism has been one of the key strategies of environmental groups in Europe, the USA and other OECD and non-OECD countries, this remains absent in contemporary China. In order to ensure being in the 'political mainstream', some local ENGOs have aligned their activities toward complementing the tasks of governmental agencies in environmental governance, such as Green Han River, a local ENGO in Hubei province.[5]

In all, the specific political context of China frames the development of environmental activism and forms an often-heard second reason why the emerging Chinese environmental movement is so different from that in other parts of the world.

Cultural context of China

The third reason brings us to the heart of this study: Chinese ENGOs are different from Western ENGOs due to the specific Chinese cultural context, which frames and structures the development, organization, and operations of environmental groups.

Chinese society has long been a small-scale rural economy, where the family was the basic unit for individuals who live by the land (Jin, 1984). These living conditions conditioned and cherished the Chinese cultural heritage. Several cultural traditions have dominated Chinese society throughout its remote history, from the Mao Communist era to its recent form, including Confucianism, Taoism, and Buddhism.

Among these traditional schools of thought, Confucianism has been the most prominent in dominating Chinese history. After being established in late Spring and Autumn (770–476 BC), Confucianism had been the ruling ideology in China for more than 2000 years, during which it was seriously revised, particularly in the Song Dynasty (960–1279 AC). Confucianism integrates a philosophy of nature with ideas on human nature, articulates methods of self-cultivation, and has a strong impact in directing individual behavior. Buddhism was originally a practicing religion, brought to China by foreign missionaries. It was most influential in the first and the twelfth centuries, when it introduced new sacred objects, symbols, buildings, and rituals, as well as new ways of thinking and interacting. To some extent it advocated principles that contradicted the core ideas of Confucianism, such as monasticism and individual spiritual enlightenment. Buddhism also introduced new ways of thinking and interacting with the material/natural world in Chinese society (Kieschnick, 2003). Taoism was founded by Li Dan, and later spread by Zhuang Zhou. This school of thought stresses *Dao* or 'way' in the universe, which is suggested to be the guiding principle that controls every aspect of the universe. Taoists sought for harmony between human beings and nature, where the focus is not on the characteristics of human beings, but on natural ones. This school became a religious sect that gained popularity and developed into a religion during the Southern and Northern Dynasties (420–589 AD).

These cultural traditions, which the Chinese have inherited from the past, are still influential in contemporary Chinese society. Their impact can be witnessed in two aspects that are relevant for this study. First, compared with Western and European countries, in everyday life in China informal rules play a significant role, whereas laws are of lesser importance. Second, the attitude toward communities in China notoriously differs from that in Europe and the USA.

First, informal rules, ethics, and customs are significant in structuring social practices and institutions in China. The authority, norms, and rules of the traditional Chinese state were prescribed by the principles of Confucianism (Duara, 1995), thus differing from Western societies, in which individuals are bound by laws and legal institutions (Landes, 1998; Metzger, 2001). Although the traditional Chinese state was a centralized unity, the local units played an important part in the governance of China, along with the institutionalized bureaucracy (Fei, 1947; Zhang, J., 2000). These local units are referred to as 'localities' (Shue, 1988). The governance of localities existed parallel to the central government, and facilitated the latter's rule. At the grassroots level of localities, social elites have great authority in their communities. Relying on informal rules, customs, and ethics, local elites played a significant role in maintaining security and coordinating local communities (Sun, 1996). These informal rules and customs worked prominently in binding individuals' behavior in everyday life in Chinese societies (Metzger, 2001). This was true already before the communist turnover in 1949, when the communist ideology and ruling were built on, rather than against, these informal rules and customs in local governance.

Second, influenced by traditional cultural ideologies, the Chinese hold specific values toward their community. The Chinese are noted to be more concerned with collective goods (Hofstede, 1983; Gudykunst, 1997). More specifically, a number of cultural principles can be identified as dominant. For instance, Chinese individuals are taught to keep public interests in mind, and those giving too much voice to particular or private interests are usually regarded as selfish. Public consensus in traditional Chinese society was supposed to be achieved by individuals sacrificing their private interests for collective goals. Individuals should prioritize responsibilities before pursuing and enjoying freedom, and they should preserve harmony and social integration in their communities and avoid conflicts. These principles significantly affect the Chinese perception and definition of human rights, which differ greatly from those in Western democratic states. This contributed to heated discussions in the international community.

The political behavior of Chinese citizens was, and is, influenced by their cultural heritages. Traditionally, individual citizens were not inclined to conduct opposing activities toward ruling authorities. Citizens are connected to the state in a hierarchical relationship, and were taught that bureaucratic institutions are designed to guarantee that bureaucrats will faithfully carry out policy made by the government (Kuhn, 1995). Although the Chinese central government has never been based on elections, the political institutions developed their own systems of political accountability to ensure that the public interests are protected (Kuhn, 1995). Also, before 1949 the Chinese relationship between individuals and state authorities differed from that in Western democracies, in which citizens obey the law while asserting their rights and standing up for democratic institutions (Pye, 1981). Traditionally, the Chinese remained subject and subservient to state authorities, and cultivated relatively little initiative in realizing their rights in political governance.

Although Chinese traditional cultural heritage had profound influence on individual behavior and attitudes towards authorities, its role was strongly affected after the establishment of the Communist Government in 1949. Chinese communism has, to a large extent, changed the (dominant) role of traditional Chinese culture in society and governance. It has been argued that during the communist era the role and influence of traditional cultures changed dramatically (Levenson, 1964; Metzger, 2001). In the early years after the 1949 turnover, traditional Chinese culture was tolerated to some extent. In the Cultural Revolution, however, all forms and expressions of traditional Chinese culture were suppressed. When Deng Xiaoping became the political leader of China in the early 1980s, the primacy and rigidity of Mao's ideology was lessened (Lieberthal, 1984; Pye, 1986). Less emphasis has been placed on the establishment of a ruling ideology. Yet essential elements in traditional Chinese cultures that could benefit the rule of the Chinese Communist Party (CCP) were adopted and celebrated. The core of the CCP strategically utilized the essential ethics and norms of Confucianism to strengthen its dominating status among Party members and toward the public (Jin, 1984). For instance, as part of the moral education under the Communist regime, the citizens of China learned to celebrate five loves: of the country, the people, work, science, and public property (Li, M., 1990). The CCP also emphasized the value of collectiveness, in line with Confucianism. As studies indicate, in comparison with the Taiwanese, people in Mainland China have a stronger preference for values related to society and country, instead of individual values (Yuan & Shen, 1998).

It is often claimed that these values of collectiveness and obedience to the ruling government have also greatly affected the formation of environmentalism in China. As the environment is regarded as a public good, protecting the environment is propagated as making contributions to a harmonious society (Pan, 2006). According to this interpretation, individuals are obligated to improve the environmental conditions of their community. Yet, at the same time, the way in which environmentalism is organized in Chinese civil society is strongly structured by the cultural heritage: non-confrontational, no direct conflicts with authorities, and strongly organized via informal channels and mechanisms. Much of this cultural specificity comes together in the role that personal networks are believed to play, both in China at large, and also in the organization and functioning of Chinese environmentalism.

Networks and environmentalism

In understanding the environmental movement, a dominant perspective nowadays is to focus on the networks in which environmental organizations and activists are involved. Network analyses are believed to provide insight into the roles and functioning of environmental interest organizations, and the rules, resource exchanges, interaction patterns, coalitions, and cultural identities of networks are the key characteristics in understanding the influence and functioning of environmental interest groups in wider social systems.

When analyzing networks in the emerging environmental movement in China, we will have to interpret these against the specific social context of the Chinese society. As noted by Diani & Donati (1999), the structure of environmental organizations and the environmental movement is not similar in every society. It is related to the specific social environment in which such movements emerge and function. Several scholars have emphasized the considerable role of personal networks in the development and functioning of Chinese ENGOs. According to them, if we want to understand the current environmental movement in China, the focus should be on personal rather than organizational networks. The leaders and key figures in contemporary Chinese ENGOs play important roles in the environmental movement by using their personal networks, connections and resources. Three—not necessarily mutually exclusive—explanations can be given for the strong emphasis on personal networks.

First, the major role of individual networks in the Chinese environmental movement follows from the specific cultural features of Chinese society. The institutionalized practice of using personal connections and networks follows particular rules and rituals, and plays an important role in social life (Yang, M., 1994; Hwang, 1987). This is not different in ENGOs.

Second, under the control of the Chinese Communist Government, the development of environmental movements has developed different strategies from those of its Western counterparts. As Zhao notes, the unique relationship between the state and society determines the action repertoire of the Chinese social movements (Zhao, D., 2005). Under the restrictive regulation of the CCP, personal connections are needed both to survive and to gain access to and influence in political and policy-making processes. The resources mobilized and strategies utilized by these organizations thus reveal differences from their Western counterparts.

Third, the strong reliance on personal networks and connections in the Chinese environmental movement relates to the relatively young character of the movement. While in the early stages of ENGO development in Western societies (in the late 1960s and 1970s), personal and individual networks were also highly relevant for movement formation, although at the present stage of environmental movement development this is no longer the case. The evolution of these Western ENGOs gradually moved them into more professionalized entities, which formed organizational networks within and beyond the environmental movement. Most Chinese ENGOs are still in their embryonic period. Lacking facilities and material and human resources, they usually grow relatively slowly, and have no formalized organizational structure. In order to mobilize resources, a large number of Chinese ENGOs thus rely strongly on the personal connections of the leader or key figure, not unlike what Western ENGOs used to do in their early development.

Research questions

To understand the phenomenon of personal connections and networks in the contemporary Chinese environmental movement, this study investigates the construction and functioning of social networks in and around environmental movement organizations. By studying the roles played by leaders, activists, and members, and their individual and organizational connections and networks, further insights can be gained on the specificity of the Chinese environmental movement. More specifically, this study focuses on the following three research questions.

- How can we characterize the specific development of Chinese ENGOs—compared with Western ENGOs—in terms of their formalization and networks?
- Are the cultural features of Chinese society, or the early stages of ENGOs development, decisive for the utilization of personal networks as their main strategy of Chinese ENGOs?
- How do Chinese ENGOs use social networks in resource mobilization, influencing state policies and the realization of their goals?

Outline of this book

In investigating the Chinese environmental movement, focusing on these three research questions, empirical field work on ENGOs has been carried out at four locations. This book reports on the research. It is structured in seven chapters. In the first chapter, I give a brief overview of the history and the roles and characteristics of contemporary Chinese ENGOs. It starts with a review of the characteristics of (environmental) governance in China, followed by the development of Chinese ENGOs, including the current administration of NGOs and their interactions with other players. In the second chapter, the conceptual framework for the whole study is developed, by first reviewing the literature on social movements and network analysis, followed by the construction of a conceptual model for studying personal networks in Chinese ENGOs. The chapter concludes with a discussion of the methodology for selecting the cases and designing the research methods. The third to sixth chapters contain four case studies on the construction, features, and functioning of personal networks in the ENGOs in four different Chinese cities: Panjin, Beijing, Xiangfan, and Shanghai. In each case, the political, economic, and cultural conditions of the city are discussed, which constitute the contexts against which ENGOs emerge and function. Subsequently, one or several ENGOs are chosen as sub-cases in each chapter to study in detail the utilization of individual networks in environmental movements. Finally, each chapter further analyses one local environmental campaign carried out by one or more ENGOs, in order to explore the strategy and resources of ENGOs, particularly regarding the relevance of personal networks. The final chapter compares these cases, provides explanations of the findings from them, and draws conclusions concerning the role of personal connections in Chinese ENGOs.

1 Environmental governance and NGOs

An introduction to China

Introduction

Following three decades of environmental governance, China is now confronted with a serious dilemma in its modernization process. On the one hand, it is experiencing fast economic growth, aiming at doubling the per capita gross domestic product (GDP) between 2000 and 2010.[1] On the other hand, its large population, high growth rate, and resource-intensive development strategy have caused serious environmental deterioration, which will continue for the next 15 years. This ecological deterioration affects continuous economic development, and environmental pollution has been a source of social unrest in the country (Economy, 2003) since the late 1990s. Since 1997, the number of conflicts over environmental issues has been escalating by 25 per cent a year (Yang, G., 2007). Recently, the Central Committee of the Chinese Communist Party (CCP) published a report acknowledging that environmental deterioration and pollution is one of the four leading causes of social unrest in the country. However, in the process of restoring the environment, the capabilities of the Chinese authorities to deal with the deteriorating environment are still perceived as weak. Confronted with the increasing amount of environmental unrest, the CCP's legitimacy of governance is increasingly being challenged.

In such a situation, the public in China is gradually starting to play a role in improving ecological governance by participating in environmental non-governmental organizations (ENGOs) and civic voluntary groups. The Chinese ENGOs have become the building blocks of the growing environmental movement in China. Due to their efforts, increased participation in environmental policies is being witnessed.

This chapter provides general information on the emergence of Chinese ENGOs in the context of the specific social and political environment in China. The next section discusses the characteristics of Chinese environmental governance, by focusing on the bureaucratic structure and the government's capacity in the field of environmental protection. Section 3 focuses on the development of Chinese ENGOs by reviewing the administrative regulations and policies on NGOs, providing a historical review on the development of ENGOs, and analyzing the organizational characteristics of contemporary environmental activism. The subsequent section explores environmental awareness in China, in order to better understand the growth (and current limitations) of green activism. The last section considers some influential actors who have a close relationship with Chinese ENGOs.

Environmental governance

In most countries, the state plays a crucial role in protecting the environment and providing environmental welfare (Xia, G. 2000; Mol, 2006). This is also valid for China. But in fulfilling its environmental responsibilities, the Chinese state differs from most other countries in its combination of a transitional economy/society, where the state still plays a crucial role in most economic and social activities, with a developmental process. This causes a number of particularities for the environmental state and environmental governance in China. Here we focus on the bureaucratic structure and the environmental capacity of the Chinese state in environmental governance.

Bureaucratic structure of environmental governance

The Chinese environmental protection organizations are located in the state's administrative structure. Environmental protection agencies are the main body in Chinese public administration. They are established at four levels of government: national, provincial, municipal and county governments. At national level, the Ministry of Environmental Protection (MEP) is funded by, and operates under, the direct leadership of the State Council. At the other three levels, Environmental Protection Bureaux (EPB) are governed by their corresponding authorities. At provincial government level and below, EPBs implement policies and laws promulgated by the MEP. As the primary goal of Chinese authorities has been rapid economic growth over the past two decades, considerable authority and resources have been vested in institutions that support economic development (Jahiel, 1998; Zheng, 2001), such as planning commissions (*Jiwei*), economic commissions (*Jingwei*), construction commissions (*Jianwei*), and so on. Environmental protection agencies have not been granted sufficient authority, according to most scholars. It is recognized that in both central and local government, the MEP and other EPBs at various levels enjoy little discretion.

Aside from the inadequate power given to EPBs, the Chinese political bureaucracy, as a result of the planned economy and the central government's strong monopoly, hampers efficient, coordinated action on environmental issues. The connection of local agencies with their functional related agencies at the center (or with a functional related agency below) is called a line (*tiao*) relationship. Line relationships represent administrative control, and governmental units have authority over each other. This authority comes only from the related agency functioning in the same line. The unit's link with other government organs within the same jurisdiction is described as an area (*kuai*) relationship. Governmental units of the same jurisdiction do not have discretionary power over others at the same level. In many cases, EPBs have to issue regulations jointly with other departments or through the mayor's office in order to enhance their stature and legitimize their regulatory activities (Lester, 1988). This bureaucratic structure suits China's multi-regional character and helps the central government to control around 30 provincial regions and a few dozen functional ministries (lines). In each region, the hierarchical structure duplicates the central government's structure. This framework suits coordination within one industrial sector, but works poorly when cross-sectoral interests are combined (Ma & Ortolano, 2000; Zheng, 2001).

More specifically, environmental governance is a complex process in which environmental agencies, various other government units, industrial bureaux, factories and even the CCP are involved (Jahiel, 1998). Yet the efforts required by government units for coordination in China are lacking in environmental governance. Conflicts and cracks exist in the administrative system, including conflicts between the upper and lower ranks, among units in the same lines and areas, or between lines and areas (Lieberthal & Lampton, 1992). What is more, the four-level governance constrains environmental agencies in utilizing their authority, as they do not have autonomy and are dependent both on their higher-ranking level EPBs and on local governments. In general, contractual stipulations are hard to achieve through coordinated efforts of government agencies, so potentially powerful allies can rarely be mobilized among environmental bureaux in order to fight against polluting establishments (Tang *et al.*, 1997). When one unit does not have complete authority, one way to put pressure on other government units is to build public consensus (Lieberthal, 1997). Consensus-building is essential to a government unit that has a low rank or authority in policy processes. This is how they strike for power balances with more powerful units in the fragmented bureaucracy (Lieberthal & Lampton, 1992).

Capacity of environmental agencies

Under the current administrative structure of the Chinese government, the capacity of environmental protection authorities is usually considered very weak. As Lindley suggests, institutional capability means the ability to choose what ends to pursue and the will to take actions to achieve them (Lindley, 1975). In this process, human and financial resources are two important aspects in environmental governance (Schwartz, 2003).

The country as a whole suffers from a shortage of both human and financial resources in environmental protection. First, lack of staff in environmental protection agencies leads to weak administration, and produces an obstacle that restrains the EPB's capabilities. Observing the pattern of staff in EPBs, we see that although the total number of staff in environmental protection agencies has increased dramatically over the last 17 years, the proportion of increase in the different levels of governmental units differs (Figure 1.1). This increase is recognized primarily at the lowest government level—county and village level (Figure 1.2). The staff in these EPBs in 2006 is more than twice the number of staff in 1994, and these two levels account for 65 per cent of all environmental staff employees in the country in 2006. However, the environmental protection agencies at provincial and city levels developed at a slower rate. Between 1994 and 2006, provincial and national EPB staff increased by 46 per cent, while staff in city government environmental protection agencies increased by 45 per cent in 13 years. This situation can be explained by the increased number of tasks in environmental management. The central government, faced with the deteriorating environment, realized the urgency of

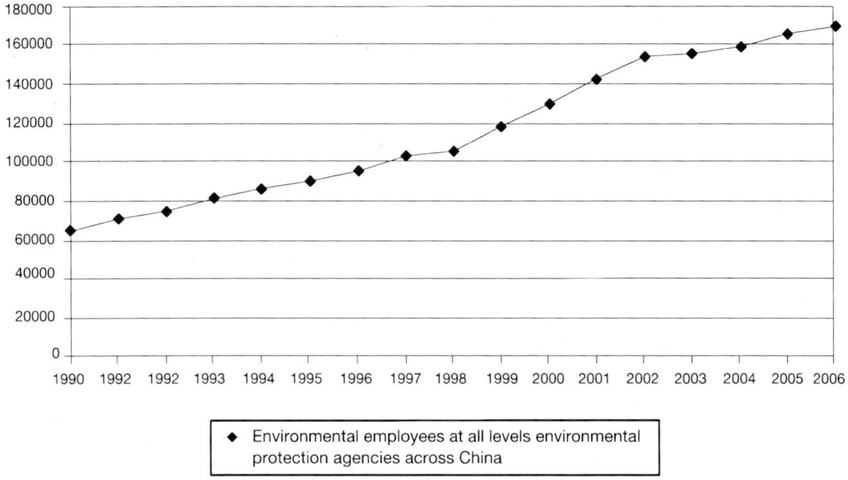

Figure 1.1 Environmental employees at all environmental protection agencies across the country

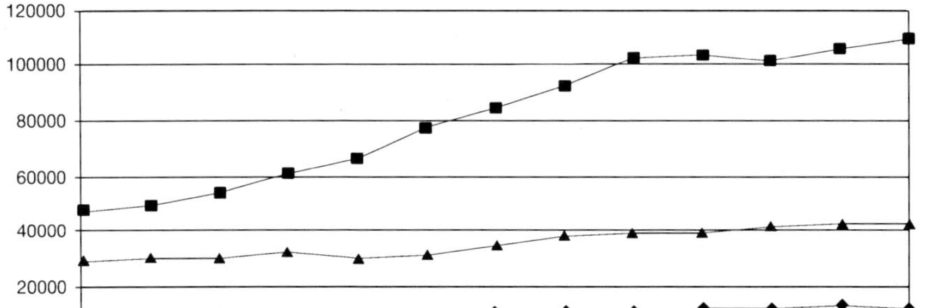

Figure 1.2 Official environmental employees at different governmental levels in China
Source: China Environment Statistical Report (1997–2007); China Statistical Yearbook (1997–2007).

establishing and strengthening authorities at the grassroots level to help clean up the environment.

Furthermore, when we look into the increased number of staff in environmental protection agencies at different administrative levels, the inspecting and monitoring departments have experienced the largest growth. At country level, the inspecting agency experienced the largest growth in all three levels of environmental protection agency, tripling in 13 years (Figure 1.3). In provincial governments, the number of inspecting departments almost doubled as compared with 1994 (Figure 1.5). In both municipal and provincial governments, the numbers of inspecting departments are now double those of 13 years ago (Figure 1.4). However, aside from the supervising departments, namely inspecting and monitoring departments, other environmental protection agencies did not witness such prominent growth in their employee numbers. Statistics show that employee numbers in environmental research institutes have hardly increased in both municipal and provincial government; only 0.9 per cent and 11 per cent more employees are recruited respectively over the last 10 years (see Figure 1.4 and Figure 1.5). Due to the administrative reform that started in the mid 1990s, there was a small reduction of officials in provincial and municipal EPBs in 1998 and 1999.

In addition, government officials at the local level generally have relatively low environmental awareness and less interest in and commitment to protecting the environment. Only a small number of environmental officials have developed a high level of expertise in their field (Schwartz, 2004). In order to supervise officials in protecting the environment, a series of documents and regulations have been promulgated by the State Council and MEP.[2] Officials thus have to

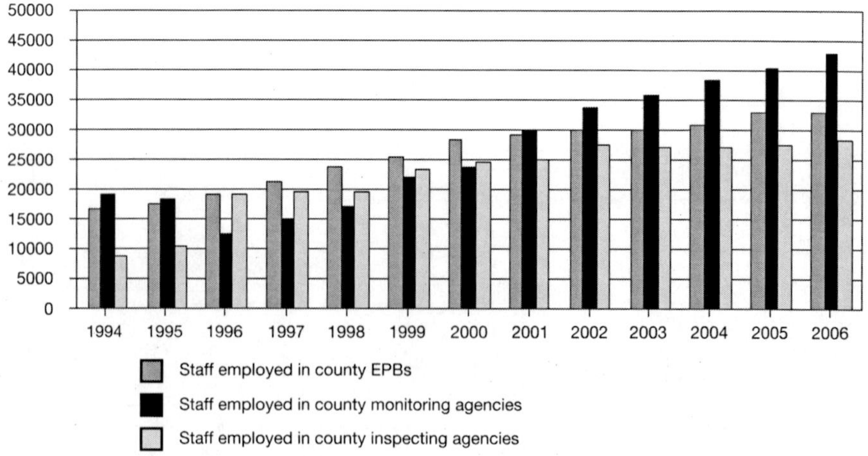

Figure 1.3 Environmental employees in county level governmental agencies (1994–2006)

Source: China Environment Statistical Report (1994–2007); China Statistical Yearbook (1994–2007).

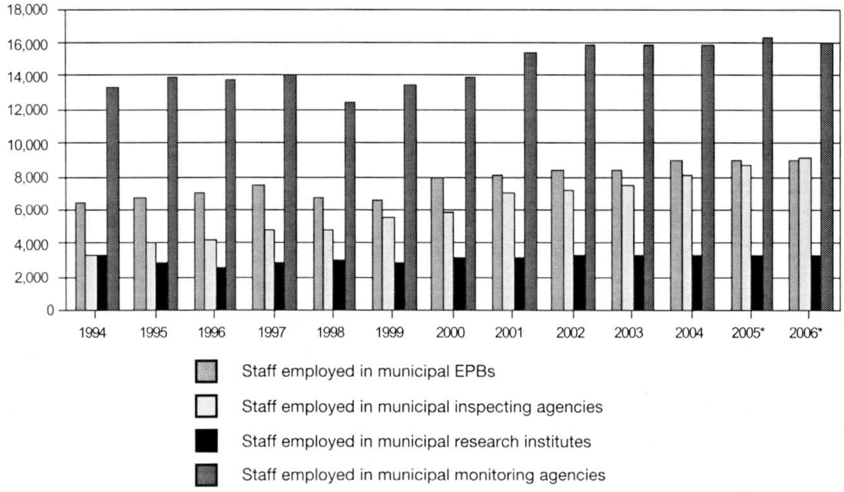

Figure 1.4 Environmental employees in municipal governmental organizations
Note: figures for 'Staff employed in municipal research institutes' in 2005 and 2006 are estimated according to historical trends. Source: China Environment Statistical Report (1994–2006).

be cautious about protecting the environment. A new recruitment assessment system, based on environmental protection performance, is being developed after having been tested in 10 cities in China.[3] Viewing the changes in different departments in environmental protection agencies, the Chinese government appears to be responding to a deteriorating environment, but ineffectively shouldering its responsibilities in ecological governance.

There is a lack of financial resources in environmental protection, particularly in local government. Along with market development, the role of the state as a provider of public goods is diminishing (Wang, S., 1997). After the fiscal reform of 1994, provincial governments became economically self-reliant: they are left with a reduced role in revenue collection, but greater responsibilities in revenue expenditure on public goods in comparison with the pre-reform era. Industrial pollution comprises one of the major funds in governments' financial resources dedicated to environmental governance. It is composed primarily of three types of fund: constructive environmental funds, which come from central government's budget; special funds that are distributed primarily by provincial and local governments; and other funds that include self-raising funds, domestic loans and foreign investment. Seen from statistics, the proportion of the funds distributed from local governments' annual budgets—special funds—has been small over the period 1996–2006 (Figure 1.6). Its proportion in the total industrial pollution treatment investment shrunk from 9 per cent in 1996 to 3 per cent in 2006. The constructive environmental funds that are distributed from the central government's budget vary significantly over the

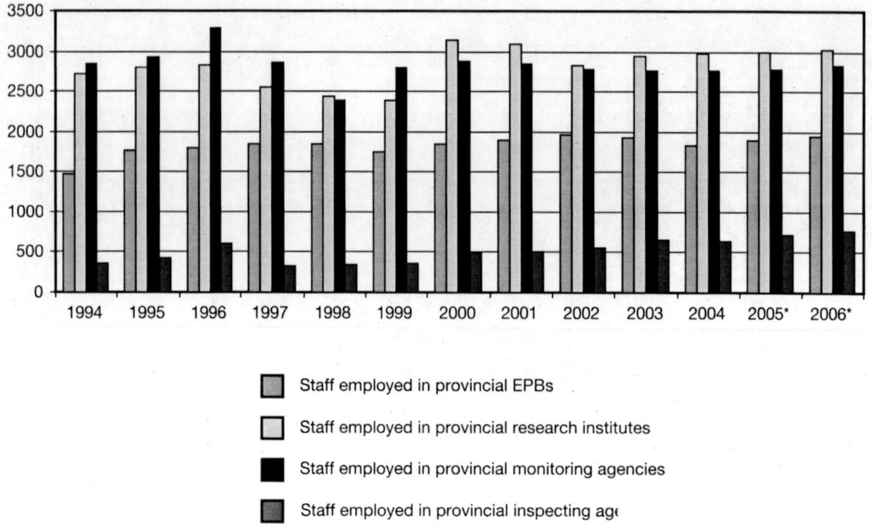

Figure 1.5 Environmental employees in provincial environmental agencies (1994–2007)
Source: China Environment Statistical Report (1994–2007).

years, and it only accounts for 3 per cent in total investment in the year 2006. Financial incomes that do not directly come from governments—self-raising funds, domestic loans and foreign investment—sources distributed outside governmental apparatuses—are the main financial sources in pollution control (Figure 1.6). Most of the local EPBs are short of revenue. As suggested by Beach (2002) and Bates (2006), in many cases, local sources of revenue are not adequate to cover costs of environmental services, and financial capacities of local EPBs are closely related to local economic development. In wealthier coastal provinces and in large cities, environmental protection tends to be better funded than in the poorer interior parts of the country, both in absolute and in relative terms (Economy, 2003). For instance, in 2006, Shanghai, the revenue for environmental protection accounts for 3.02 per cent of local GDP; while in the Sichuan province of the west, no more than 0.3 per cent of local GDP is invested in environmental protection. Since provincial governments have the authority to direct EPBs, which the national level parent ministries lack (see, for example, Economy, 2003; Beach, 2002), the provincial shortage of resources makes local governments relatively weak in environmental management.

Increasing revenues have been invested merely to keep the environmental situation from deteriorating further (Wang, L. *et al.*, 2004). Between 1996 and 2000, 3600 billion RMB was spent on pollution prevention, and this figure has doubled to reach 8288 billion in the five years since 2001 (White Paper, 2006). In 2006 alone, 1.22 per cent of the total GDP was spent on environmental pollution enforcement. Yet such investment does not cover the damage caused by environmental pollution. According to a report on Assessment of Green GDP (2004), made jointly by the MEP and National Bureau of Statistics,

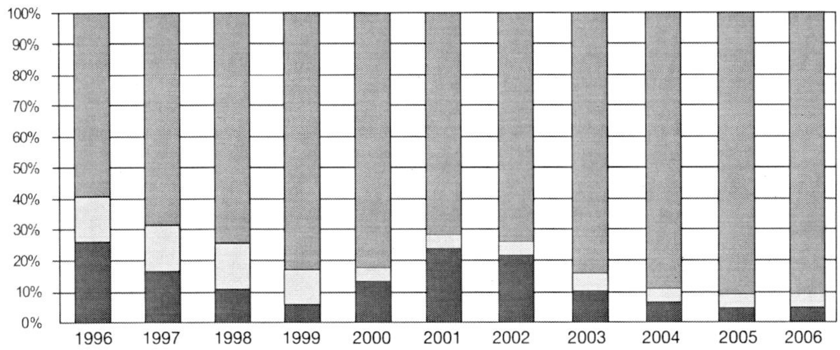

Figure 1.6 Sources of current investment in the treatment of industrial pollution (1996–2006)
Source: China Environment Yearbook 1997–2007.

economic loss through environmental pollution accounted for 5118 billion yuan in 2004, which accounted for 3.05 per cent of the GDP that year. Of the total economic loss, water pollution, air pollution and solid waste constitute the major categories in environmental compensation, which account for 55.9, 42.9 and 1.2 per cent, respectively.[4]

In the evaluation of the national environmental performance, the Chinese government ranks 94th among 133 countries, and ranks only 15th among the 17 Asia–Pacific Economic Cooperation (APEC) member countries (Yale Center and CIESIN, 2006). Among the six main indicators that are used to measure government performance, Chinese environmental governance is below average in biodiversity and habitat policies, in comparison with the average level of its peer countries in the same income group and geographical group. It scores lower than the average of its peer group in the indicators of environmental health, productive natural resources and sustainable energy. China's policy on air quality and water resources scores particularly low in comparison with its peer group. This comparative approach has its limitations in measuring the effectiveness of the state's environmental policies, because the performance of the state cannot be measured merely by the environmental performance of a country, which is affected by a variety of other social causes (Duit, 2005). Nevertheless, the current Chinese bureaucracy has difficulties in managing environmental protection. As argued by Jahiel, the Chinese environmental protection institutional setting has limitations in effectively cleaning the environment (Jahiel, 1997, 1998).

Environmental NGOs in China

Against this background of environmental governance, green activism has been emerging from the early 1990s onwards in China. Along with it came a growing desire to establish self-organized voluntary organizations among the public. The two trends resulted in the establishment of ENGOs in China. Being the first and most active type of NGO, ENGOs are active in representing the public in Chinese environmental governance. Established in relatively repressive legal and political circumstances, the Chinese environmental organizations have encountered a great number of regulations and restrictions (section 1). Nevertheless, as reported in the following section, their organizational development and relatively late emergence distinguish current Chinese ENGOs from those flourishing at the moment in Western democratic societies (section 3).

Government policy on NGOs

In recent years, China has witnessed an upsurge of NGOs (Wang & He, 2004). These private and voluntary social organizations are defined here, following the definition for civil society of Salamon & Anheier (1997), by the following five criteria.

- Organized—they have some structure and regularity in their operations, whether or not they are formally constituted or legally registered.
- Private—they are not part of the apparatus of the state, even though they may receive support from government sources.
- Not profit-distributing—they are not primarily commercial in purpose and do not distribute profits to a set of directors, stockholders, or managers. While NGOs may generate a surplus from time to time, they must reinvest these resources back into the objectives of their respective organizations.
- Self-governing—they have their own mechanisms for internal governance, instead of being dominated by political authorities.
- Voluntary—membership or participation in them is not legally required or otherwise compulsory.

This study targets ENGOs that commit to improving the environment. More specifically, I focus on the types of environmental group that are interested in mobilizing civic activism and are concerned with civil society's development. This feature distinguishes them as NGOs that have providing services as their goal. In this study, ENGOs are alternatively referred to as voluntary groups or civic organizations.

In general, the Chinese authorities are not keen on encouraging organized efforts by the public, which are often associated with political dissent. The state has established restrictive administrative procedures concerning NGOs. The Ministry of Civil Affairs and its regional bureaux are the official bodies responsible for management and regulation of voluntary groups. NGOs in China generally include three types.

- Social organizations (*shehui tuanti*)—this type of organization includes associations, societies, federations and research associations.
- Foundations—social organizations that raise funds, or distribute or operate activities based on funds. The purpose of these organizations is primarily

for commonwealth or public charities, for instance the China Environmental Protection Foundation and the China Foundation for Property Alleviation.
- Non-profit enterprises—social organizations that provide social welfare and public service, such as schools, hospitals, etc.

Chinese NGOs began to increase rapidly at the end of the 1980s, and reached their peak around 1996, when 186,666 registered NGOs existed. However, the regulations of the state prevented their further growth, and a considerable number were closed after the collapse of the Falun Gong practice. In 2003, 142,000 officially registered NGOs remained (Wang & He, 2004).

At present, the state's regulation of the NGO sector is characterized by strict registration procedures. In 1989, the State Council promulgated the 'Regulation on the Registration and Administration of Associations'. According to this law, the authoritarian state leaves extremely little latitude for self-organized social groups and their involvement in politics and policy-making.[5] With some exceptions, each association has to first find a party/state agency that is willing to be its official sponsor.[6] Failing to do this would result in termination. In addition to the regulations imposed by its supervising institutions, NGOs are also under inspection of the Ministry of Civil Affairs, which implements regular administrative surveys on the management of non-profit organizations, in compliance with the legal framework of the NGO sector. This system of dual regulation was designed for the party-state to control the development of the registered associations. It thus becomes difficult for social organizations to set up a bank account for public donations.[7] Under this management system, ENGOs usually have only a limited degree of autonomy. Also, initial funds are required for registering social organizations.[8] In this way, the authorities attempt to have further control over the growing number of social groups.

As an alternative to being registered as a social organization, some NGOs have been registered as a company. In this case, NGOs are treated as for-profit entities and are therefore in a higher tax bracket. In practice, the remittance is changing due to the accounting reforms that began in 2003. In the past, the tax authorities might have applied an informal standard that was more lax than the taxation regime for genuine for-profit corporations.[9] After 2003, it became more difficult for tax authorities to use their own discretion to 'turn a blind eye' and not insist on full payment of corporate tax. Registered corporations are required to pay tax monthly on all income received. For Chinese NGOs with little financial support, this policy will prevent them from registering as companies. Thus a larger number of environmental groups continue to exist as unregistered NGOs.

Historical review of ENGOs

Although certain environmental groups existed before 1990, most non-profit environmental organizations at that time were government-organized NGOs (GONGOs). Independent ENGOs were established from the early 1990s onwards.

A limited number of ENGOs were established in the early 1990s, and an increase in their numbers was seen in the mid-1990s (Figure 1.7). This process was greatly affected by the globalization process, as a host of international events integrated China further into the world community, expanding domestic political arenas and boosting environmentalism (Yang, G., 2004). Chinese environmentalists were particularly enlightened by the Rio Declaration following the UN Conference on Environment and Development that was held in 1992. During the same period, multilateral banks, bilateral aid agencies and international NGOs (INGOs) poured into China, bringing funding, projects, expertise and legitimacy to Chinese ENGOs. After the UN World Conference on Women, held in Beijing in 1995, Chinese citizens were greatly enlightened by the international NGO culture, and many Chinese citizens began to initiate their own voluntary groups and organizations. As environmental protection is an issue of relatively low political sensitivity, the first environmental groups were formed.[10]

The number of ENGOs has grown rapidly in the past ten years. According to a survey, by the end of 2005 there were approximately 2768 ENGOs (ACEF, 2006).[11] The growth of the Chinese ENGOs sector is estimated around 10–15 per cent annually (in terms of both number of organizations and number of staff working for them). Within this sector, 49.9 per cent are GONGOs, 47.5 per cent are ENGOs (among which students organizations account for 40.3 per cent); and 2.6 per cent are INGOs (Figure 1.8). Yet this result does not provide specific data on unregistered ENGOs, such as web-based organizations and ENGOs registered as business organizations.[12] According to an estimate, the total number of unregistered environmental organizations is over 2000 (Economy, 2005).

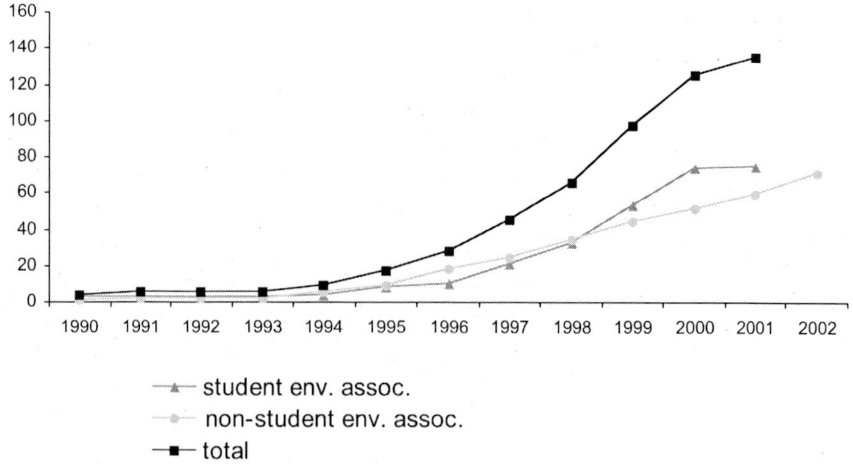

Figure 1.7 The development of student and non-student environmental associations (1990–2002)

Source: Yang, G. (2005: 51).

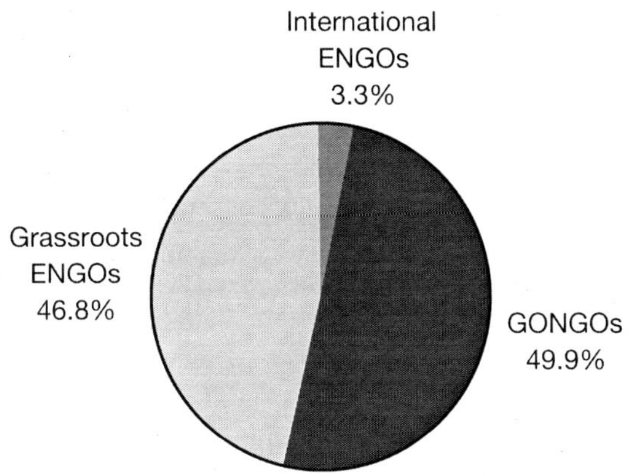

Figure 1.8 Percentages of GONGOs, ENGOs and international NGOs in 2005
Source: ACEF (2006).

Regarding the location of ENGOs, they have developed in several focused regions across China. According to the survey conducted by All-China Environmental Federation (ACEF), currently ENGOs flourish mainly in four areas: Beijing and Tianjin in the north; Shanghai on the east coast; the west southern area of Sichuan and Yunnan provinces; and Hunan and Hubei in the middle provinces, where ecological resources are relatively rich (ACEF, 2006). In Beijing, pre-existing personal networks facilitate the establishment of ENGOs and the organization of environmental activities. The activists are mostly members of the social elite, well known to each other, and anxious about the political reform of the Party and the development of China. Some of the activists participated in the 1989 Student Movement; some were connected with Dai Qing, an environmental activist who became famous for her protest articles on the Three Gorges Dam.[13] Establishing an ENGO is one way of realizing their social ideals. Guided by prior experiences in the previous movement, these *avant-gardists* are skillful in coping with the repressive government. Tianjin is geographically very close to Beijing, and therefore is easily influenced by the ENGO community that prospers there. On the east coast, where economic prosperity is high, individuals have relatively more free time to commit to voluntary activities, so private, voluntary groups are active in this area. In the middle provinces, environmentalists focus on the protection of the relatively rich ecological resources (ACEF, 2006). The provinces in west southern China have witnessed the establishment of international NGOs since the early 1990s. Local green activism was strongly enhanced by the funds and knowledge they contributed.[14] Being close to Bangkok—headquarters of many international NGOs in Asia—also helped local Yunnan NGOs to build international liaisons.[15] Local environmentalism was thus fostered in this province.

Characteristics of contemporary ENGOs

Chinese NGOs and informal volunteerism have been most active in the area of environmental protection, and take diverse organizational forms. There are six main types of independent ENGO:

- NGOs registered as social organizations or private, non-profit work units (*minfei*)
- non-profit enterprises registered as business enterprises but operating as non-profit organizations
- university research centers/institutions affiliated with institutions of higher learning that operate as NGOs
- student environmental associations registered with campus Youth Leagues, yet functioning and perceived as NGOs
- voluntary organizations that function as NGOs
- web-based groups operating mainly via the internet (Table 1.1) (Wang & He, 2004; Yang, G., 2005).

As noted by scholars of the Chinese environmental movement, Chinese environmental activists have a mild attitude toward the government, based on their avoidance of challenging state authority (Saich, 2000; Martens, 2006). ENGOs' stance in adopting non-critical views towards government is evident in the activities they conduct. At present, the majority of the groups are committed to environmental education as their primary activity (Turner, 2003; Yang, G., 2005). Some ENGOs, particularly research institutes or centers, focus specifically on one particular issue, such as environment-related public health problems, pollution control, biodiversity preservation and environmental education; whereas others, especially voluntary associations, tackle a broad range of concerns, from tree planting through dam construction to energy conservation. Most ENGOs are preferably engaged in raising environmental awareness of citizens and officials, in disseminating information to the wider public, and in clean-up operations in streets and parks. In recent years, an increasing number have become engaged in lobbying central government on issues such as promoting recycling or wind power, or bringing to a halt large-scale dam construction.

Regarding the organizational development of green activism, the Chinese independent ENGOs usually have limited resources. They lack financial resources, office space and supplies, and skilled personnel (Table 1.2). These problems are obstacles to the prosperity of these ENGOs (Turner, 2003), and for most,

Table 1.1 Types of registered and unregistered environmental NGO

Type of ENGO	*Registration status*	*Example*
Voluntary associations	Registered as social organizations (*shehui tuanti*) or private, non-profit work units (*minfei*)	Grassroots Community (Shanghai), Green Han River (Hubei); Green River
Non-profit enterprises	Registered as business enterprises	Institute of Environment and Development (Beijing), Green Cross (Beijing)
Research centers /institutions affiliated with institutions of higher learning	Not registered with Ministry of Civil Affairs	Center for Legal Assistance to Pollution Victims (Beijing) (registered with China University of Political Science and Law); Global Environmental Institute (Beijing) (registered with Chinese Academy of Sciences)
Student environmental associations	Registered with campus Youth Leagues	City Group of Shanghai Normal University (Shanghai), Green Wild (Xiamen)
Web-based groups	Unregistered	Green Web (Beijing), WWF Shanghai (Shanghai)
Voluntary organizations that function as NGOs	Unregistered	Green Earth Volunteers (Beijing); Green Friend Association (Shijia Zhuang)

the shortage of funding poses a serious problem (Schwartz, 2004; Wang, A., 2005). Domestic fundraising is impeded by an absence of government funding for citizens' initiatives and by restrictions on raising money and donations. ENGOs receive hardly any financial support directly from government organizations. What is more, under the current legal framework Chinese donors receive negligible tax cuts when donating to social groups. Such unfavorable tax regulations largely reduce the chances of ENGOs receiving donations.

So, presently, most ENGOs are small and have difficulties in finding financial support due to their limited knowledge of fundraising. The majority of them manage on an operational budget of less than 100,000 RMB (approximately US$15,000); and around 80 per cent of these groups spend no more than 50,000 RMB (approximately US$7500 $) annually (ACEF, 2006). Only a small number of ENGOs work with a budget over 100,000 RMB (Deng, 2001). Aside from financial problems, NGOs also suffer from a lack of facilities (Deng, 2001; Turner, 2003; ACEF, 2006). Sixty per cent of ENGOs do not have their own offices. Around half have computers in working order and can afford to publish a newsletter. Lack of personnel is one of the main difficulties. The average scale of each group is small, many having no full-time staff. Around 44 per cent of full-time staff do not receive a salary. In groups that pay their

Table 1.2 Major problems encountered by ENGOs

Problem	*Percentage of ENGOs*
Lack of funding	41.4
Lack of office space and supplies	11.7
Lack of qualified personnel	9.9
Internal management problems	7.5

Source: Deng (2001).

staff regularly, 96 per cent of the remuneration was approximately at the mid- or lower level of local salaries. Over 50 per cent of staff do not have any social security (ACEF, 2006).[16] Owing to this situation, the majority of ENGOs lack environmental expertise and professional knowledge (Deng, 2006).

In addition to a lack of financial resources, environmental organizations are confronted with serious problems in organizational development (Deng, 2001; Turner, 2003), and have limited knowledge about how to manage their groups professionally. A large number of China's NGO leaders lack management skills and have no effective governance structure (Zhu, 2004; Qian, 2006). For instance, a number of ENGOs do not have a board of directors, or have one in name only. So, as Young (2001) suggests, Chinese ENGOs greatly need to foster transparency and accountability in their internal governance, in order to increase their organizational capacity.

Although still developing slowly, the establishment and the growth of Chinese ENGOs are the result of efforts made by a great number of enthusiastic environmentalists. China's environmental activists are composed of a relatively homogeneous population, who tend to be educated and, in many cases, quite charismatic. About 80 per cent of them are less than 30 years old; 70 per cent of the leaders are less than 40 years old. Over 90 per cent of their leaders have received higher education (ACEF, 2006). The first generation of environmental activists was made up of scholars, journalists and intellectuals, who were concerned about environmental deterioration and committed themselves to realizing social responsibilities (Fu, 2004). This group of social elites is capable of mobilizing resources such as funds, contacts and information. Some of them are well known figures with social leverage. Young people also constitute another group of environmental activists. According to the report of ACEF, of the 224,000 ENGO staff (including full-time and part-time staff) who were surveyed, 80 per cent are below 30 years of age (ACEF, 2006). As followers, the young people intend to gain social experiences and realize self-fulfillment through voluntary work (Yang, G., 2005).

The Chinese environmental activists committed to environmental protection have shown solidarity in their activism. In many ENGOs, because of the lack of financial resources, staff are not provided with proper social welfare (ACEF, 2006).[17] According to the ACEF report, 91.7 per cent of the members who work in environmental groups are not paid any allowance. Commitment is the primary basis for individual environmentalists' participation. Their activities thus assume a high level of volunteerism. Also, the idea of improving the environment builds solidarity among individual environmentalists (Ho, 2001). In this way, despite the lack of formality within Chinese ENGOs, these groups can survive and successfully mobilize activism.

Environmental awareness in China

The emergence of ENGOs is closely related to the growing environmental awareness of the Chinese public. In recent years, the public has been very concerned about environmental problems. The escalating environmental deterioration, an increase in environmental reporting and higher living standards contribute to the rising environmental awareness of Chinese citizens.

The environmental consciousness of the public is made up of their awareness of environmental problems and their participatory behavior. In general, the Chinese are very concerned about environmental protection, evident from their perception of environmental problems and their willingness to participate in activities protecting the environment.

Seen from the number of letters of complaint and visits to environmental protection agencies pertaining to environmental problems, it can be concluded that over ten years the total number has increased (Table 1.3). Statistics show that among the representative sample of 10,495 households interviewed in 2000 (from 31 provinces, autonomous zones and municipalities), 57 per cent believed

Table 1.3 Statistics for environmental pollution accidents and number of complaints

	Number of environmental pollution accidents	*Compensation for environmental pollution or loss (yuan)*	*Number of letters of complaint*	*Number of visits to complain*
1995	1966	3,854.7	58,678	94,798 (persons)
1996	/	/	67,268	47,714 (times)
1997	/	/	106,210	71,528 (persons) 29,677 (times)
1998	/	/	147,630	493,791 (persons) 40,151 (times)
1999	/	/	230,346	89,872 (persons) 38,246 (times)
2000	/	/	247,741	139,424 (persons) 62,059 (times)
2001	1842	12,272.4	369,712	80,575
2002	1921	4,640.9	435,420	90,746
2003	1843	3,374.9	525,988	85,028
2004	1441	3,487.2	595,852	86,317
2005	1406	2,373.8	608,245	88,237
2006	842	7,396.5	616,122	71,287

Source: China Environment Statistical Reports (1995–2003).

that environmental problems in China are 'very serious' or 'relatively serious'. Only 23 per cent believed that the environmental problems are 'not too serious' or 'can be ignored' (Xi & Xu, 1999). Yet, despite a general acknowledgement of environmental problems, the public's perception of specific environmental problems is rather limited, evident in their knowledge on specific environmental issues (Hong, 2005; Minsheng Bank, 2005). The most distressing environmental issues include noise pollution, and air and water quality. This is understandable as these problems can be sensed directly by the public (Minsheng Bank, 2005). Yet, in comparison with their perception of daily issues, the public's concern about regional and global environmental issues is relatively low. Only 0.8 per cent of the public expressed knowledge of sustainable development (Minsheng Bank, 2005). Around 41 per cent of interviewees had never heard of 'acid rain', while only 5 per cent of them were well aware of this issue (Hong, 2005). Thus Chinese citizens display different perceptions of local, regional and global environmental issues (Figure 1.9).

Further, statistics have illustrated that people are more inclined to participate in environmental actions if their action is linked to clear benefits and visible results. According to the research of Hong (2005), conducted in 2003 with 5073 respondents, a relatively low degree of environmental participation exists in China, measured by indicators such as avoiding the use of plastic bags, paying attention to news of environmental protection in the mass

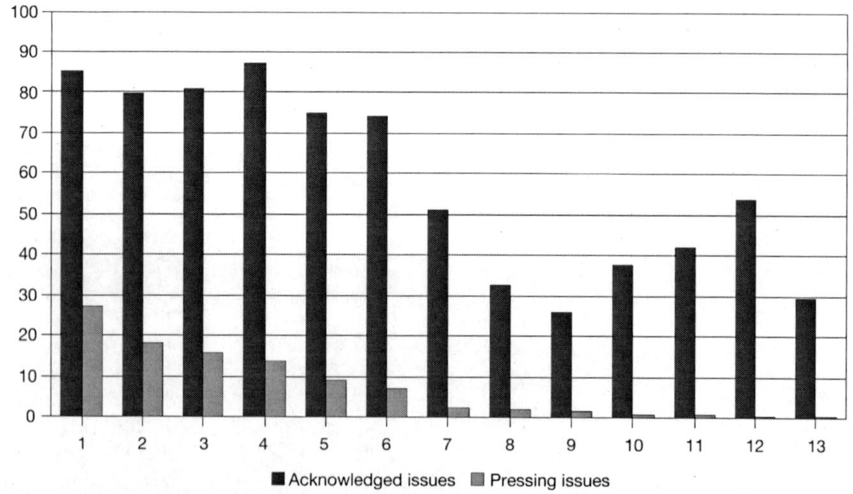

Figure 1.9 Knowledge of the Chinese public on environmental issues (percentages)
Source: Minsheng Bank (2005).

Notes: 1 General environmental pollution; 2 air quality; 3 solid waste management; 4 waste water management; 5 city greening; 6 health problems; 7 tidiness of cities; 8 environmental degradation; 9 energy issues; 10 nature reservations; 11 sustainable development; 12 non-renewable energy issues; 13 biodiversity preservation.

media, donating for environmental protection, planting trees, and so on. High levels of participation in environmental protection are relatively non-existent. For instance, around 76.1 per cent of interviewees had never participated in activities organized by ENGOs (Table 1.4). However, 48 per cent of respondents indicated that they probably will participate in environmental protection activities organized by ENGOs, and a further 46 per cent indicated that they may not get involved—as shown by a survey conducted among 3777 respondents in 20 cities across China (Figure 1.10).

According to the survey carried out by Hong, the level of concern on local environmental issues correlates with people's income and educational level (Hong, 2005). In the poorest group, with an income of less than 200 yuan per month, 25 per cent of respondents show a very low level of environmental concern. Yet in the most affluent group of citizens, only 12.4 per cent have the same low level of environmental awareness (Table 1.5). This result leads

Table 1.4 Participation in environmental protection activities

Activities	*Never (%)*	*Sometimes (%)*	*Often (%)*	*No answer (%)*
Separate waste	62.9	21.7	15.1	0.4
Discuss environmental problems with friends	31.6	56.6	11.6	0.2
Avoid using plastic bags	49.6	27.5	22.6	0.3
Re-use plastic bags	29.0	24.6	46.1	0.3
Donate for environmental protection	69.1	27.3	3.2	0.4
Attentive to environmental protection news in mass media	23.0	45.1	31.7	0.2
Participate in government- or *danwei**-organized environmental protection activities	58.7	30.0	10.9	0.4
Participate in activities of ENGOs	76.1	18.5	5.0	0.4
Plant trees voluntarily	82.0	12.0	5.6	0.4
Actively involved in lawsuits concerning environmental rights	82.6	14.5	2.6	0.3

Source: Hong, 2005.

Notes:*In China, *danwei* refers generally to a special mechanism under the centrally planned economy. It provides welfare to carry out government policies, ranging from family planning to public hygiene campaigns. *Danwei* is also a social institution that provides for employees' welfare benefits such as healthcare, housing, daycare and pensions. It also works as a social and political control mechanism for the Chinese government. This phenomenon is becoming less common nowadays after the introduction of the market economy.

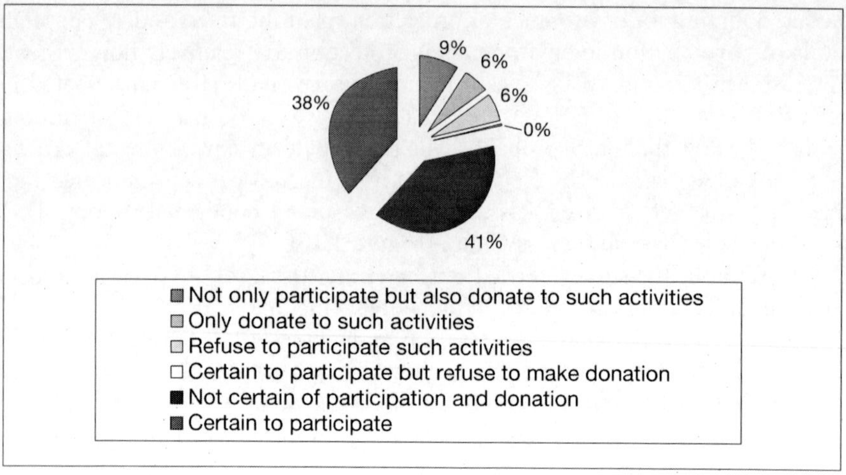

Figure 1.10 Public participation in environmental protection activities and donations
Source: Minsheng Bank (2005).

Table 1.5 Scores for environmental knowledge and individual income levels

Score for environmental knowledge	*Persons with income level (%)*				
	<200 yuan	*200–500 yuan*	*500–800 yuan*	*800–1,200 yuan*	*>1,200 yuan*
<50	25	23.3	18	17.4	12.4
50–60	36.2	37.2	33.8	31.8	24.0
60–70	22.2	24.5	28.2	25.8	30.0
70–80	12.2	10.9	13.2	18.4	20.7
>80	4.4	4.2	6.0	6.6	12.9
Total	100	100	100	100	100

Source: Hong (2005).

indirectly to the conclusion that affluent people have relatively high environmental awareness compared with poorer people, when measured by the indicators used.[18] According to the same survey, over half of the interviewees who had a bachelor degree scored over 70 for their environmental knowledge, while over 70 per cent of interviewees who had only a preliminary level of education scored less than 60 (Hong, 2005) (Table 1.6). Thus level of education has a linear correlation with the level of environmental awareness when measured in this way.

When exploring further the degree of environmental commitment in relation to citizen behavior, researchers have found that actual support for environmental protection is particularly low when citizens are confronted with hard choices (Lee, Y., 2005; Minsheng Bank, 2005). For instance, a survey conducted in Beijing and Shanghai produced empirical evidence that the degree of public

Table 1.6 Scores for environmental knowledge and educational level

Score for environmental knowledge	*Persons at educational level (%)*				
	Preliminary education	*Middle school*	*High school*	*Certificate*	*Higher than bachelor degree*
<50	34.4	23.2	15.4	9.8	7.2
50–60	39.8	38.0	32.4	23.5	15.8
60–70	18.3	25.0	28.6	30.6	24.9
70–80	5.9	9.8	17.7	20.6	32.4
>80	1.6	4.0	5.9	15.5	19.8
Total	100	100	100	100	100

Source: Hong (2005).

commitment towards environmental protection was somewhat lower when people were asked to pay extra fees for environmental friendly products and services (Yuan, 1999). The degree of commitment dropped significantly when respondents were asked to make major adjustments in their consumption behavior that might compromise their quality of life.

The relatively low level of environmental participation, as opposed to the still significant concern for environmental problems, is usually explained by individuals' preference of participating only when behavioral changes are linked to clear benefits and visible results (World Bank, 1996).

Actors who influence Chinese ENGOs

Through the environmental campaigns of recent years, several actors have made an important contribution to the emergence of ENGOs and the collective actions they organized. Government organizations, international NGOs, media and private companies constitute four important categories of actor.

Government and GONGOs

Before 1992, the Chinese government undertook almost all environmental public affairs, and the public had little opportunity to participate. Since the emergence of ENGOs in the early 1990s, the Chinese public began to some extent to articulate their interests in ecological governance. Yet the current legislative institutions provide limited access for individuals to become involved in environmental protection, apart from environmental pollution law cases.[19] This situation in environmental management greatly diverts from the requirement of China's Agenda 21 that advocates comprehensive policy-making and cooperation in environmental protection (Zheng, 2001).

Lately, faced with serious environmental deterioration, the Chinese government has begun to show a less firm attitude towards single-handed environmental governance. Relying solely on the government's power in this process requires large amounts of money, adding further to the cost of environmental governance, and worsening the situation of limited financial resources for enforcement of sanctions (Zhang & Xia, 2000). The government can scarcely afford the tremendous work of environmental protection.

Due to this situation, the central government has started adopting a positive attitude towards ENGOs. Incorporating public and voluntary groups in environmental policy-making has been emphasized by central government in a series of documents and regulations. In the Law on Environmental Assessment that was promulgated in 2003, assessments for constructing projects are subject to public consultation. This law helps institutionalize the inclusion of public comments in environmental decision-making (Turner, 2003), and has empowered environmentalists in their activism. The state has begun to provide openings for ENGOs in environmental governance.[20] This legal framework is complemented by 'Environmental Protection Administrative Licensing Hearings Provisional Measures' (2004) and 'Provisional Measures for Public Participation in Environmental Impact Assessment' (2006). Both legal documents delineated measures for citizens' participation in environmental impact assessment (Moore & Warren, 2006). The emphasis placed on this issue by central government indicates that a certain role is available for ENGOs, which is expected to supplement the state's capacity to improve the environment effectively.

Besides this, the Chinese government has also initiated cooperation with the public and ENGOs. In this area, a special type of NGO, GONGOs, play a key role. According to the survey conducted by ACEF, the number of GONGOs in China is now 1382, including different types of foundations and associations (Figure 1.8). Examples of GONGOs that are committed to environmental protection are the China Environmental Protection Foundation (established in 1993), and the All-China Environment Federation (established in 2005).

GONGOs differ from ENGOs in many aspects, but primarily in their relationship with government. They usually have a large number of government officials serving as GONGO staff or on the board of directors (Knup, 1997). In many cases, the reasons for establishing these institutions are to create more jobs (Turner, 2003), and these organizations are often funded by government. Because of their organizational characteristics, the establishment of GONGOs is often suggested to be beneficial to retired officials, who may be recruited as leaders. And the level of expertise in many of these organizations is impeded by a lack of professional knowledge and techniques.[21] Despite the dependence of GONGOs on the government, these organizations represent a potential source of independent environmental activism, initiated by government agencies or institutions (Wu, 2002a; Schwartz, 2004).

Regarding the relationship between GONGOs and ENGOs, both competition and cooperation exist. Compared with harsh competition between the two actors in other fields, particularly philanthropic fundraising,[22] in the field of environmental education cooperation is seen more often between these two types of social entity, and they often implement projects together. GONGOs provide ENGOs with networks to access policy-makers and state and economic institutions. Most importantly, voluntary environmentalists can implement their projects by using the administrative structures of GONGOs (Zhao, X., 2004). Otherwise, ENGOs could not have been accepted as legitimate actors to conduct environmental protection projects, particularly in remote areas where public perception of NGOs is relatively low. The participation of GONGOs to some extent brings legitimacy to ENGOs' projects, which facilitates the latter interacting with the public and local political authorities in launching their projects. ENGOs, at the same time, bring GONGOs useful information and more project opportunities. ENGOs can act more quickly than GONGOs in response to the needs of the public in environmental protection. At the same time, they are capable of launching projects, and a number of ENGOs show efficiency in mobilizing volunteers and expertise in environmental education. Through cooperation, the two actors complement each other and succeed in improving environmental reforms outside the state.

It can be predicted that in the future a level of competition will arise between the two actors. However, as environmental degradation remains serious, and the state's capacity relatively weak, the two actors see more opportunities to cooperate than to compete with each other.

International institutions

Two decades ago, a number of international NGOs began to enter China. To date, it is estimated that there are over 200 INGOs operating in China, including those from Hong Kong.[23] Among these, some focus on fostering the development of civil society, for instance the Ford Foundation, the Asia Foundation, and so on. Around 23 per cent of INGOs are environmental organizations that focus on environment and biodiversity in China,[24] for example WWF, Conservation International, the Nature Conservancy, the International Fund for Animal Welfare, and so on.

INGOs have proven to be a strong force in fostering the development of Chinese ENGOs, especially when the latter were just established in the early stages of NGO development. They bring significant funding to help other Chinese groups. These funds reach billions of dollars, and constitute the major source of revenue for Chinese ENGOs (Fu, 2004; Yang, G., 2005).

Aside from financial provision, interaction with international NGOs facilitates the professional development of Chinese environmental groups. As the management of non-profit organizations is relatively new to Chinese citizens, they have obtained their knowledge of management and project monitoring almost entirely from INGOs.[25] In cooperation projects between these two players, the Chinese partners often imitate the deeds of their counterparts. Particularly when receiving funding, Chinese NGOs have to establish structured working systems such as planning, evaluation, and (financial) reporting systems before the INGO will become involved with and/or fund a project.[26] Through these processes, Chinese NGOs have learned how to manage their programs in a formalized way. INGOs also organize seminars and workshops to train Chinese NGO staff, in order to improve their skills and techniques in fund-raising and project evaluation.

Furthermore, INGOs assist Chinese ENGOs to stay informed of developments in the world through conferences and visits, and provide opportunities for environmental activists to communicate with ENGOs in developed countries. With the organizational development of domestic environmental groups, such interactions are continued between the two counterparts. In recent years, an increasing number of international groups and foundations have established themselves in China, and various forms of cooperation have been identified (Turner, 2003). As a partner of a Chinese ENGO, an INGO is likely to facilitate the latter's development in the long term.

Mass media

In addition to INGOs, the mass media are another force in developing and shaping the environmental movement. In recent years, they have played a significant role in raising public environmental awareness and disseminating the activities and points of view of ENGOs to a wider public.

Chinese mass media enjoy a certain freedom in reporting environmental issues. Unlike those in western countries, several main media groups, such as Xinhua News Agency, Chinese Central Television and China Central Radio Station exert significant power in determining which stories receive coverage, and what approach journalists should take when reporting a story. They also serve as news providers for the entire Chinese media structure (Wen, B., 2003). While there is still considerable state control on the media, it is no longer monopolistic. Specifically, environmental issues have been given more freedom in the press and on television (Shambaugh, 2000; Haggelund, 2004). At local level, environmental reporters claim that they have more leeway on these issues than on other topics, through which they establish influence on local government (Turner, 2003).

With a higher degree of freedom in environmental issues, the media yield a steadily increasing number of reports in this field since the early 1990s. One national NGO, Friends of Nature, traced the yearly number of environmental articles in major national and local newspapers in China from 1994–99. Between 1997 and 1999, the number of articles on the environment doubled. While 76 newspapers produced 22,066 articles on the environment in 1997, 75 newspapers produced 47,273 environmental articles in 1999 (Friends of Nature, 2000). Among these articles, the percentage of in-depth reporting (investigations, featured articles, editorials) accounted for an average of 20 per cent. Many Chinese TV stations now have regular environmental educational programs, and there are a growing number of radio programs, environmental hotline call-in shows and exposés on local pollution violations. Internal reference memos are also frequently sent to Party leaders by key activists to cover protests or demonstrations on environmental issues that have been given less attention in the public media.

The internet is another major form of mass media utilized by environmental activists. The 1990s saw dramatic growth of an internet population and culture in China. The percentage of internet usage is high. Official reports indicate that almost half the population in Beijing and Shanghai 'frequently use internet', indicating that reading the news is one of the chief reasons to use it.[27] Of the numerous civic organizations, China's environmentalists were among the first to embrace the internet to advance their cause. Very often, ENGOs are found to have at least a minimal level of internet proficiency, and they utilize the internet in a variety of ways: email is the most frequently used by Chinese civic associations, followed by search engines and websites. Electronic newsletters and bulletin board systems are often utilized as well (Yang, G., 2007). Thus environmental activists have taken advantage of this new method of communication to disseminate information and networking with peer and international organizations (*ibid.*). With the help of the internet, ENGOs can establish a network that has significance in China's political context.

Yet freedom to access the internet is under the control of the Chinese government. The authoritarian administration of the internet is relatively repressive: it closes internet sites, intimidates actual and potential users, and imposes repressive policies on internet service providers. In recent years, the government has exerted extensive power and resources and found increasingly sophisticated methods of controlling internet usage. Websites and bulletin board forums of ENGOs are thus not free from government surveillance and control (Yang, G., 2007). The development of environmental activism that is facilitated by the internet has relatively limited freedom.

Private companies

Business actors comprise the last important category that deserves examination in environmental governance. In recent years, helped by the globalization and internationalization of trade and investment, Chinese companies and enterprises began to be aware of their accountability in improving environmental protection in China. Although their interaction with ENGOs remains limited, they are potentially interesting actors in the development of the environmental movement in China.

Private companies have demonstrated their support for the activities of ENGOs primarily by providing them with financial resources. An increasing number of enterprises show enthusiasm to donate or provide funding for ENGOs. However, the current legal framework does not provide incentives for this kind of support. The Chinese government has issued very limited initiatives to encourage enterprises to act charitably towards NGOs (Deng, 2006). The Public Donations law, adopted in 1999, is the first legal document regulating donation activities in China, and merely defines a basic framework for charity donations. Its implementation has been difficult due to its vague nature and inconsistencies with other regulations. For example, according to Articles 24 and 25, business enterprises and individuals enjoy tax breaks when donating money to public welfare undertakings. However, the exact amount of tax benefits and the way in which enterprises can benefit are not specified.

Chinese local enterprises lack accountability, and thus do not play a prominent role in Chinese ecological governance. Being corporate citizens, local companies are widely recognized as failing to fulfill their responsibilities in environmental protection (Institute of Environment and Development, 2005). In this regard, bilateral or international companies' initiatives differ greatly from local players, acting as partners working with local communities in environmental protection (Turner, 2004). They are known to be relatively active in fostering local environmental activism. For instance, in Shanghai, international companies have become partners with local ENGOs in conducting environmental education to the public (Zhou, 2004). They benefit from such cooperation by improving their relations with local communities, and improving their public image. However, domestic enterprises seldom have such initiatives.

As argued by Mol, because they lack autonomy under the current political system, economic actors are not motivated to incorporate environmental interests in their operation; and have not played any significant role in environmental reforms (Mol, 2006). Their activities in collaboration with ENGOs are also relatively limited.

2 Analyzing Chinese environmental movement networks

Introduction: defining social movements

In the literature on social movements, the concept of social movement varies according to different schools of thought. In the past several decades, this phenomenon has been defined primarily by its organizational form and social consequences. Several main definitions of social movement were put forward in early social movement studies.

McCarthy & Zald (1977) regard most mobilization efforts as organized by associations or collectivities of associations. They use the term 'social movement organizations' (SMOs) to refer to activist groups possessing the skills and contacts necessary to mobilize grassroots actors in pursuit of causes. Similarly, Tarrow (1983: 7) sees 'social protest movements' as 'groups possessing a purposive organization, whose leaders identify their goals with the preference of an immobilized constituency which they attempt to mobilize in direct action in relation to a target of influence in the political system.'

Social movements were also defined by their political outcomes. Tilly defines social movements as a series of contentious performances, displays and campaigns by which ordinary people made collective claims on others (Tilly, 2004). For Tilly, social movements are a major vehicle for ordinary people's participation in politics. Concerning the relationship between social movement and politics, it is emphasized that clear boundaries exist between social movement and institutionalized politics (Tarrow, 1988): movement actors' participation is achieved through public protest (Rucht, 1990, 1995).

Today, the revision of the social movement concept is reflected in a stronger emphasis on ideological themes and a shift away from focusing solely on the political orientation of movement actors. Recognizing the links and ties between the various SMOs and the significance of collective actions, Della Porta and Diani (2006: 20) emphasize three features of this phenomenon.

1 It is involved in conflictual relationships with clearly identified opponents. This aspect refers to the importance of broadened patterns of change that

are increasingly recognized as movement consequences (Oliver *et al.*, 2003; Della Porta and Diani, 2006).

2 Social movements are linked by dense informal networks. A social movement is argued to be 'loosely-structured' (Oberschall, 1980), and described as 'a network of informal interactions between a plurality of individuals, groups and/or organizations' (Diani, 1992).

3 A distinctive collective identity. A plurality of concepts, ideas and beliefs can now be identified when investigating the composition of social movements (Tilly, 1998). This aspect highlights the role of ideological themes in collective actions.

In this study, the environmental movement is defined as civil society-based networks of organizations and individuals that share a collective identity of environmentalism. In other words, the environmental movement is characterized by two features: environmentalism as the collective identity that binds actors together; and collective and often coordinated actions to improve environmental quality. A social movement consists of the ties and webs created to enable people to act collectively. So the movement network functions to facilitate mobilization, concrete coordination and cooperation. Consequently, the environmental movement is defined by the sharing of environmental or green collective identities, often through networking activities to improve environmental quality.

In this chapter I develop a conceptual framework within which to study and analyze environmental movement networks in China. First, I focus on reviewing the various social movement theories (section 2); then I investigate literature on China's contemporary social movement (section 3) before reviewing networking approaches in movement studies (section 4) to build a conceptual framework for classifying and understanding the specifics and diversity of environmental movement organizations (section 5), with specific reference to the Chinese network society (section 6). The latter two sections are devoted to a reformulation of the research questions and the design of the research methodology.

Social movement theories: an overview

Usually, four main approaches are distinguished in contemporary social movement studies. They provide different perspectives for examining social movement dynamics. The resource mobilization theory was developed primarily in the USA around the 1960s and 1970s; the new social movement approach is a more western European approach of the 1970s and 1980s; political opportunity structure theory has been developed primarily by European scholars since the 1980s; social constructivist theories saw their development both in the USA and in Europe in the 1990s. Reviewing the strengths and weaknesses of these approaches is a first step in developing a theoretical framework for investigating networks in the environmental movement in China.

Resource mobilization theory

Resource mobilization theory has been a dominant theory in social movement studies for a long time. It derives from organizational studies, and thus highly values the functioning of formal SMOs. This approach provides significant insights into how movement actors develop strategies and interact with their environment in order to pursue their interests.

The resource mobilization approach is very effective in analyzing mobilization processes of social movements (Klandermans, 1991), seeing the mobilization of resources as the key process that defines the social movement. Researchers in this tradition have focused on two aspects of organizational analysis: the interaction of SMOs with their environment, and the organizational infrastructure that supports social movement activity.

In the first aspect, the concept of resources is believed to be crucial in organizing the movement. This approach presumes that rational actors engage in instrumental action to secure resources needed. If the necessary resources are acquired, a social movement is more likely to achieve its goals. The concept of 'resource' has been defined rather differently as the theory has evolved. Early resource mobilization theory was ambiguous as to the definition of 'resources' and its implications for movement development. In the early 1980s, the term was criticized as 'primitive' and 'unspecified' (Gamson *et al.*, 1982). According to early resource mobilization theory, resources that are valuable to social movements can include: manpower (including time); money; expertise; contacts (and networks); legitimacy; and facilities (McCarthy & Zald, 1977). An expansion of the resource mobilization model is to stress the necessity of mobilizing time as an important resource as well as money (Rosenstone & Hansen, 1993). When studying the viability of homeless SMOs in the USA, exploring a detailed typology of potential resources, Cress & Snow (1996) developed a list of resources for sustaining collective efforts in a social movement. It includes: moral resources (referring to the endorsements, legitimacy and moral support that come from other organizations); material resources (not only money, but also what money buys—supplies, meeting space, office space and transportation); information resources (referring to those that bring knowledge and skills to a group); and human resources (referring to the people who populate groups, associations, organizations and movements).

Resources have a crucial influence on the development and success of collective actions. The way and the extent to which movements acquire these resources from their constituencies shape their activities. The availability of resources affects the rate and spread of protest (Minkoff, 1997; Soule *et al.*, 1999). It also determines where movement mobilization happens. Metropolitan regions and large countries where potential activists are able to access a wide range of resources usually recognize the emergence of a social movement (Lincoln, 1978; McCarthy & Zald, 1988). Transnational SMOs are usually found to grow largely in wealthy democratic countries (Smith and Wiest,

2003). Moreover, the amount and diversity of resources have differential influences on SMOs' survival. The more diverse resources a SMO has, the more autonomous it is, and the more likely it is to be adaptable to challenges and capable of survival (Alexander, 1998; Ganz, 2000).

In the second aspect, organizational theory has been applied within this approach by identifying organizational evolution and organizational structure. It is argued that formal organizations are carriers of social movements (Zald & McCarthy, 1987: 12). In mobilizing resources, they compete with other organizations pursuing similar goals. Resource mobilization theory suggests inter-organizational competition is essential to movement activity (Koopmans, 1993; Minkoff, 1994). Regarding organizational structure, formal SMOs promote the stability and continuity of social movements (Staggenborg, 1988; Minkoff, 2001) as well as movement survival (Walker & McCarthy, 2007). Leaders of formal SMOs are viewed as skillful political entrepreneurs who play a crucial role in organization maintenance. They found organizations, and respond to incentives and risks in the external environment, and the strategic decisions they make also influence inter-organizational cooperation and the formation of coalitions (Staggenborg, 1986; Zald & McCarthy, 1987; Morris & Staggenborg, 2004). Members of SMOs are seen rather as rational actors supporting leaders and as less important in organizational life (McCarthy *et al.*, 1973, 1977; Oberschall, 1973).

Resource mobilization theory has little to offer as to why social movements are formed, and is also weak in analyzing the consequences of social movement actions. Yet this approach is powerful in explaining the making of social movements. Its focus on organization-level processes provides insights to assess actors' activities and the survival of social movements. It ignores the impact of contextual conditions on the making of social movements (Kriesi, 1991). This approach assumes that the challenge for social movements is to identify and seize opportunities for action, which is based on a cost–benefit assessment of the likelihood of success. Therefore social movements can be investigated by assessing factors that affect SMOs' survival, such as leadership, organizational formality and organizational structure (Minkoff, 1997, 1999; Olzak & Ryo, 2007).

As correctly concluded by Zald (1991), this perspective 'retains its ability to suggest new lines of research in both micro and macro levels of collective action generation'. It can also be extended to analyzing a social movement's links with its context, by specifying types of potential resources for SMOs. Different types of resource, and the importance of each resource, vary according to the context of social movements. In industrial democracies, movement activities are related to the democratic political conditions, where they are legitimized without being regarded as anti-system-directed. Therefore valuable resources might be concentrated in the private sector. But in constrained political systems where opposition towards authorities is restrained, a differentiation of valuable resources would be needed to affect policy-making. By comparing the differentiation of various potential resources crucial to social movements in both contexts, it is possible to find the influence of institutional, social and cultural factors on movement mobilization. Therefore this school of theory will be most useful for exploring movement characteristics if it examines the resource types that have a strong influence on specific social conditions as well as the organizational processes of SMOs.

New social movement theory

New social movement theory derives mainly from the experience of social movements in western Europe since the early 1970s (Kriesi, 1996). Issues of the environment, women's rights, peace and nuclear arms were identified as the agendas of new social movements.

In new social movements, individual activists respond in both a reactive and a proactive way to the post-modern world, resisting the negative consequences of industrialization for personal needs. In this theory, a shift in value systems is found to be of significance to movement participation, such as from material welfare and safety towards post-materialist values, quality of life, conservation of nature (Brechin & Kempton 1994), anti-growth, libertarian and populist themes, etc. (Dalton & Kuechler, 1990). Theorists emphasize that this aspect of new social movements poses a contrast to the demands of older social movements, when conflicts over material resources occurred. Ronald Inglehart was among the earliest scholars who remarked on the shift in values that took place in the post-industrial societies in Europe. He labeled these changes a 'silent revolution', and defined values that do not revolve around issues of economic security as post-materialist values (Inglehart, 1971).

The constitution of new social movements is different from that of conventional social movements. Unlike in old social movements that are based on the working class, new social movement literature recognizes a shift to a diffuse base of popular support. A number of scholars assert that new social movements are composed mainly of 'new middle class' support and professionals employed in social and cultural services (Cotgrove & Duff, 1980; Alber, 1985), and differ fundamentally from the old social movements that are constituted of the traditional labor class (Kriesi, 1988; Della Porta & Diani, 2006). This class is usually highly educated and possessing technical and cultural competence; therefore they are likely to engage themselves in fighting against agencies of late capitalist societies.

According to this approach, the collective search for identity is essential to movement formation. Theorists argue that collective identity formation is a dialectical part of collective action. The constructing process of a collective identity sees a struggle for new meanings (Cohen, 1985), which is regarded as a self-identifying process. Melucci even regards identity formation as itself the goal of a social movement (Melucci, 1989).

In this approach, the organizational structures of new social movements are defined in contrast to the organizational structures of conventional movements. First of all, the former are said to consist of 'diversified and autonomous units' (Melucci, 1996), forming a fluid structure where the borders and morphology of movement organizations constantly change (Diani, 1990). Moreover, in contrast to conventional social movements in which organizations are characterized by hierarchy and centralization (Van der Heijden, 1997), the protest groups of the new social movements are characterized by decentralized networks of groups with direct participation of the people involved. As demonstrated by Kriesi, studying new social movements in the Netherlands, informal organizational approaches reduced the formality of social organizations in the peace movement in 1980s. Such organizations have neither formal membership nor a hierarchical internal structure, which makes them deliberately minimalist in terms of organizational structure (Kriesi, 1987). The movement has a tendency to have an informal character: there is decentralization in the power distribution among movement organizations. Similarly, movement activities are suggested to be invisible, 'agglomerations of individuals, groups, events, texts, media and so on rather than as hierarchically constituted organizations' (Melucci, 1989; Gusfield, 1994; Mueller, 1994). Melucci argues that, in a sense, the latency or everyday form of movement activities is the goal of public protest (Melucci, 1989). What is more, the traditional structure of the new social movement may be changed further following the utilization of computer-mediated media, virtual media and communication technologies (Diani, 2000).

The theory has been developed largely against the background of the Western new social movements, related to a modern, developed and open society. It analyzes links between movements and their context, offering important insights into the origin and nature of contemporary social conflict. However, as concluded by Buechler (1995), its strength is also its constraint. This theory is based on the western new social movements, related to advanced capitalist, post-modern information societies. Therefore insights from new social movement theory cannot be applied and used directly in understanding and studying social movements in China, where social development is largely different from any idea of post-industrial societies.

Social construction

More recently, social constructionist scholars have shown new perspectives and frameworks for understanding social movements. Framing analysis is an approach that explores the construction of meaning and its attributes to social conflicts. By providing a theoretical bridge between social–psychological and resource mobilization approaches (Snow *et al.*, 1986; Benford & Snow, 2000), this approach looks into the ongoing process of cultural framing, the relationships between frames and related cultural phenomena, and their effects for social movements.

The core issue of 'framing' is defined as 'a process of ordinary people making sense of public issues' (Benford, 1994). The target of framing is not only 'the world in which social movements are acting', but also social problems (Snow *et al.*, 1986). At the same time, concluded by Della Porta and Diani, framing can be a merely a process of symbolic re-elaboration as well as a reaffirmation of tradition (Della Porta & Diani, 2006). Meanings are attributed to movement followers' actions and the issues— this is referred to as 'frame alignment'. Four major forms of frame alignment are frame amplification, frame bridging, frame extension and frame transformation. Frame amplification refers to 'the clarification and invigoration of an interpretive frame that bears on a particular issue, problem or set of events' (Snow *et al.*, 1986: 469). This is the most basic form of frame alignment. Frame bridging is the 'linkage of two or more ideologically congruent but structurally unconnected frames regarding a particular issue or problem' (*ibid.*: 467). It involves the linkage of a movement to 'unmobilized sentiment pools or public opinion preference clusters' of people who share similar views or grievances, but who lack an organizational base. Frame extensions are a movement's effort to incorporate participants by extending the boundaries of the proposed frame to include or encompass the views, interests, or sentiments of targeted groups. The main goal of frame extension is to make the frame more attractive to its potential adherents with different priorities. Frame transformation is a process where 'new values may have to be planted and nurtured, old meanings or understandings jettisoned, and erroneous beliefs or "misframings" reframed' (Tarrow, 1992: 188).

In this approach, theoretical extensions have been identified. In earlier times, attention was paid to a micro-level of movement participants. Its explanatory power lies best in explaining how individuals are mobilized and recruited; how a movement develops; and how collective actions are sustained (Snow & Benford, 1988: 198; McAdam, 1996). The framing process is regarded as a 'sustained and self-conscious' process that takes place among movement actors (Gamson & Meyer, 1996: 283). Movement leaders act as movement 'articulators', who largely determine the consistency of the frame, the frame's empirical credibility, and the personal salience of the frame for potential participants (Benford & Snow, 2000: 619–622). In recent years, the framing perspective has also been broadened in its conceptualization of frames as organizational properties (Snow, 2004). The framing process has proved of significance in analyzing meso-organizations in social movements, in the mobilization of groups (Gerhards & Rucht, 1995; Snow, 2004); and in neutralizing or disparaging counter-framings between movement and countermovement (McCaffrey and Keys, 2000). As generalized by Oliver and her collaborators, this approach enables us to examine empirically the process through which a given objective situation is defined and experienced (Oliver *et al.*, 2003).

In recent scholarly discussion of the framing perspective, greater emphasis is placed on framing practices and on the processes through which frames are transformed, instead of frames *per se* (Snow, 2004; Della Porta & Diani, 2006). Particular attention has been paid to the relationship between framing processes and mass media and opportunity structures.

Media play a central role in the construction of meaning and the reproduction of culture (Neidhardt & Rucht, 1991; Noakes & Hank, 2005). Media discourse, according to Gamson, is not only a tool for analyzing cultural meaning, but also is one of the focal points in which social movements wage their contests to change public policy or perceptions (Gamson, 1995). The media's nature is of great significance to the framing process. It is argued to be instrumental in advancing social movements' causes. Newspaper and television stories about a movement can raise the public's consciousness of particular problems (Gamson & Modigliani, 1989). The media also raise the profile of activists and expand their influence (Ruzza & Schmidtke, 1993; Gamson, 1995). With the media's positive discourse, a movement's collective action frame tends to have a stronger impact on a broader range of people (Walgrave & Manssens, 2005). Therefore Gamson and Meyer argue that the mass media are a privileged provider of reference frames and can serve as possible channels for diagnostic framing (telling people what the problem is) and prognostic framing (what possibly can be done) (Gamson & Meyer, 1996). At the same time, the media possess a certain autonomy and may have complicated relationship with social movements. Gamson & Wolfsfeld (1993) depict the interaction between social movement and the media: how movement contended with media to have the latter transmitting their messages, while media preferred maintaining and reproducing the dominant mainstream frames.

When analyzing the character and influence of movement framing processes, the political context in which such processes occur is also gradually considered and investigated (cf. Snow, 2004; Kriesi, 2004). Framing activities are shown to be affected by the changing political opportunities. Schneider demonstrated that the mobilizing frames and trajectories of neighborhood movements in New York City are determined by differences in the local political opportunity structure (Schneider, 2005). In turn, identities framed in movement activism are incorporated in local political culture, which serves the development of new movement frames. Formatted by political opportunities, the case of Russia also illustrates that the content and priority of framing are affected by the political conditions in which such processes occur (Zdravomyslova, 1996).

In this study, this approach provides insights for studying cultural meanings in social movements. It is helpful in explaining how values and beliefs correlate to cultural frames, and are used in environmentalist mobilization in China. The present Chinese culture is still heavily influenced by symbols and values of cultural tradition, which contain beneficial aspects for environmental protection. Therefore the Chinese movement mobilization can hardly be understood without analyzing how traditional Chinese values and beliefs are successfully confirmed or modified in producing impact on public opinion. Moreover, as the state has permitted certain free space for environmental groups, particularly those working in the field of environment (Saich, 2000), environmental activists apparently enjoy more latitude in their activities, including framing environmental activism. Although movement actors have difficulties in competing with existing legitimate frames, and are constrained by the political system in which the Chinese environmental movement is located, with the media and use of other technology the Chinese environmentalists have gained an opportunity to discuss and introduce new issues into the main official discourse.

Political opportunity structure theory

A final important theory of collective action is the theory of political opportunity structures (POS). This school of thought focuses on the relationship between political environment and social movements. It asserts that changes in political opportunities and constraints create important incentives and possibilities for initiating new phases of contention and movement activism. This perspective has proved helpful in explaining several aspects of the dynamics of social movements, such as movement strategies (Kriesi, 1991), action repertoires (Kitschelt, 1986), and the development of movements (Tarrow, 1994: 85–86; Van der Heijden, 2006).

The concept of POS was largely a contribution from political science to the study of social movements (cf. Van der Heijden, 1997). This core concept was raised in the 1970s to become influential in social movement research much later. But the conceptualization of 'political opportunity' varies greatly through time and among the major contributors. Eisinger first mentioned the idea of POS, to distinguish the different degrees of openness that individual local governments provide for participation (Eisinger, 1973). Tarrow's work provided more concrete ideas to define POS. According to him, four salient changes in opportunity structures define the potential for social actions: the access to and participation in political processes; shifts in ruling alignments; the availability of influential allies; and cleavages within and among elites (Tarrow, 1994: 85–86). Yet inconsistencies have been identified in Tarrow's concept of POS: first it was identified as an 'incentive' (1994), later as a 'signal' for social movements (1996). Thus POS has been defined and applied differently in social movement studies over time (Meyer & Minkoff, 2004). Only in the 1990s, when European scholars started to work on POS, which was elaborated and modified, was a more coherent and systematic structure of POS formed along with its application to western European countries.

Opportunities for social movements have been related to political systems primarily in two respects: the formal institutional structure of the state, and the informal strategies of political elites (Kriesi, 1992, 1995; Van der Heijden, 1997). The degree of openness of the formal institutional structure is regarded as the core variable of POS. It is further elaborated by three aspects: centralized versus decentralized state system between the national, regional and local levels; the power relations between the legislative, the executive and the judiciary; and the nature of the electoral system—majority system versus proportional representation (Kitschelt, 1986; Kriesi, 1995; Rucht, 1996). It is argued that the greater the number of actors who share political power (the greater the checks and balances), the greater the chance that social movements gain access to the system (Della Porta & Diani, 2006: 202). The informal structure of power relations is distinguished as two basic sets of strategies that political elites adopt: repression (exclusion) and facilitation (assimilation) (Kriesi, 1992, 1995; Tarrow, 1994).

Political opportunity structure theory has been developed to capture the dynamic development of political systems. The above two dimensions are relatively stable to represent political opportunity structures, and are complemented by another two variables. A third variable is 'configuration of power', which refers to the question whether the governing elite of a political system is basically divided or united. The more divided a political system is, more opportunities are provided to movement actors. 'Political output structure' is raised as a fourth variable. Kitschelt (1986) demonstrated that strong political output structures (centralized state apparatus, government control over market participants) provide fewer opportunities for movement actors than weak political output structures (decentralized state, few grip on market participants).

Political opportunity structure makes possible our understanding of the context in which activism takes place, especially at the national political level (Kriesi, 1995). As nation-states are increasingly incorporated in the international level of governance, this theory is also applied to the international level (Van der Heijden, 2006). More specifically, the POS provides a framework for analyzing the dynamics of social movements. By linking institutional properties to social movement dynamics, this approach is strong in explaining the emergence of movements (Della Porta & Diani, 2006). It is also powerful to explain levels of movement mobilizations (Kitschelt, 1986) and movement strategies (Kriesi, 1991). A number of longitudinal and cross-sectional studies illustrate that the repertoire of popular protests shifts along with the shifts in opportunity structure (Tilly, 1995; Koopmans, 1993). However, the properties of formal institutional structure are suspected to be rather stable, and a less evident factor in explaining the development of protests (Della Porta, 2005; Walgrave & Rucht, 2007). Moreover, it has to be noted that the explanatory power of the theory in this aspect also lies partly in the characteristics of popular protests that are allowed by the democracies, although the levels of mobilization differ according to specific POS. Changes of new popular contention can almost certainly be predicted. Largely depending on the novelty of movement (Koopmans, 1993), strategies can be confrontational (sit-ins, blockades, etc.), or increasingly militant or compelling increasing numbers of movement actors. From this perspective, POS's effect on movement's level and forms is very dependent on its environment. In addition, this linkage between movement strategies and political opportunity structures is based on the assumption that the movement is primarily politically oriented (Goodwin & Jasper, 1999). Other types of social movement—'identity-oriented', 'subcultural' or 'countercultural' movements—may not depend on political opportunity structures to an equal extent.

China's authoritarian context determines that its POS and movement dynamics basically differ from those in liberal democratic societies. The four extended dimensions of POS provide insights when identifying the POS of China. In particular, the informal strategies of political elites comprise an important aspect for examining the movement opportunities that are provided by the Chinese political system. There are two dimensions in this respect: integrative (distinguished by facilitation, assimilation and cooperation) and exclusive (repression, polarization and confrontation) (Van der Heijden, 2006).

In a one-party state such as China, the role of top leaders of the Communist party is extremely important. Moreover, China has developed its own cultural heritage, which affects political elites' strategies toward social movements. As the authoritarian state is increasingly inclined to accommodate movement actors in recent years, it can be recognized that certain movement opportunities are provided to movement actors. In short, the political opportunity structure theory is very relevant to analyzing social movements in China. In order for this theory to have more explanatory power, further theoretical expansion and Chinese contextualization should be made.

Shortcomings

Generally, two problems exist when investigating social movements in Southern countries. First, the existing literature views social movements as revolutionary and opposed to the regime (Boudreau, 1996: 181). In Southern countries, overthrowing or initiating political reform is usually not the primary goal of collective actions. In constrained political conditions in Southern countries, to capture resources or to shield society from state initiatives are often identified as movement efforts (Boudreau, 1996).

Moreover, movement development and strategies in Southern countries also differ from those in Northern countries. Tilly's work illustrates that variations in political and social context influence collective movement strategies (Tilly, 1993). In Southern countries, state development is still in (a modernization) process, where social and political systems are under constant reform. Social movements could hardly develop in the same ways as we found in liberal democratic societies (Neidhardt & Rucht, 1991). Instead, a variety of strategies and tactics may prove successful in reaching movement goals.

In reviewing the dominant theories and literature on social movements, it can be concluded that some of these schools can be relevant to provide insights into the development of the Chinese environmental movement. Yet each has its own shortcomings for exploring the Chinese environmental movement's dynamics. Specifically, there are two aspects of the existing theories that fail in relation to China. First, according to resource mobilization theory, SMOs have a tendency to develop into formalized organizations. At the same time, the role these SMOs play is presumed to be dependent on the social context in which they are located. Do we now see in a different context (China, with its specific culture and political system) similar characteristics of formalization in the environmental movement? Second, how do the environmental movement actors in China establish and sustain collective identities? In particular, what mechanisms are utilized in this process? Although social construction theory emphasizes the significance of collective identity in the formation of collective action, it provides little theoretical insight into how this process is organized. In new social movement theory, identity formation is related to the perceptions and ideas prevalent in post-industrial societies; but how does that work in developing and transitional countries? Do we witness similar paths and mechanisms?

Social network approaches and perspectives are less 'closed' than the aforementioned social movement theories. As a structural approach, network analyses recognize that a plurality of actors is involved in movement processes (Diani, 1992). Before exploring the usefulness of network approaches and perspectives in studying the development and dynamics of the emerging Chinese environmental movement, I will first review the existing literature on China's social movements.

Contemporary social movement in China

In the past three decades, China's economic and social transition has been a source of grievance and agitation among the public. Discontent has developed into various types of protest, ranging from peaceful resistance to radical action. Since the early 1990s, China has witnessed the growth of charitable organizations and associations 'concerned with societal groups that are marginalized and vulnerable in the process of reform'. Women's groups, legal counseling centers for women, children and environmental victims, rural development centers and organizations, and poverty alleviation associations (Howell, 2003) have all contributed to an embryonic public sphere under the rule of the Party state.

Environmental protection has been one field in which certain space is provided by the state for NGOs to establish in (Saich, 2000), so it witnesses public activism ranging from voluntary groups to online activism. Individual case studies on the environmental movement have been conducted on environmental campaigns (Zhu & Ho, 2008; Sun & Zhao, 2008); individual environmental NGOs (Economy, 2004; Lee, S., 2006) and certain groups of environmentalists (e.g. Stalley & Yang, 2006). However, 'at the vanguard of non-governmental activity' (Economy, 2004) ENGOs and environmental groups struggle to survive, being faced with an uncertain relationship with the state, financial restraints and a lack of publicity (Ho, 2001; Martens, 2006). Activists quickly embraced and learned to exploit the opportunities afforded by the internet. Studies have observed high visibility of environmentalism in cyberspace and asserted that the information and communication technologies play a role in the emergence of social movements in China (Yang, G., 2005a, 2005b). However, the scope of activities of online environmental groups remains unexplored (Ma, 2006; Yang, G., 2007).

Environmental NGOs are found to be playing an increasingly visible role in China's environmental politics. They have been affected by two distinctive actors in their evolution. GONGOs have played a significant role in bridging NGOs with government departments and coordinating with them based on mutual needs and understanding (Wu, 2002a; Lee, S., 2006). Another group playing an active role in China's growing environmental activism is the international NGOs. They assist Chinese ENGOs in financial and organizational capacity-building, and have established increasing density of linkages with local NGOs (Morton, 2008; Zusman & Turner, 2005). Overall, in spite of an increasing level of scholarly interest, our current understanding of the Chinese environmental groups is limited. Among the studies of environmental activism in Asian countries (e.g. So & Lee, 1999), the existing literature provides a limited, in-depth study of the nature, characteristics, strategies, organizational modes, and influences of what could be labeled a Chinese environmental movement in the making.

Aside from the activities recognized in associations, in the past few years a growing literature deals with China's rising social protests and instances of collective action. This literature has explored thoroughly the causes, forms of action, outcomes and interplay between state and society in popular contention in China (Perry, 2001; O'Brien, 2002; Cai, 2006; O'Brien & Li, 2006).

When exploring the causes of collective action in China, one distinctive tendency in the existing literature is to focus on the tension between state and society. However, opposition towards the non-liberal Chinese regime does not exactly constitute the source of resistance in China (O'Brien & Li, 2006). Two distinct types of protest have been recognized: first, agitation and grievances toward official misconduct. Rural areas, in particular, see large numbers of cases in this cause, for example, villagers rising when cadres violate popular notions of equity, fairness or justice (Bianco, 2001; Unger, 2002; Bernstein & Lü, 2003). In urban areas, cases are most frequent where the existing order is threatened. For instance, land requisition for development zones (Cai, 2007) and residents' increasing environmental complaints (Brettell, 2008; Zhu & Ho, 2008). A second type of popular protest is rooted in unfavorable policies in China's economic transition. For instance, workers' interests have been affected in various ways through the reform period, for example of state-owned enterprises, redundancies, pensions and so on (Cai, 2006; Lee, C., 2007).

Popular contention in contemporary China has its own characteristics, compared with that found in democratic societies and other transitional countries. The goal is seldom to overthrow the authoritarian regime. Equally, the aims of popular protest seldom involve demands for wider political change. Very few instances have challenged the regime (Cai, 2006; O'Brien & Li, 2006). The largest number of popular protests indicates resistance to the authorities within the boundaries of authorized channels. They appear as 'non-contentious' episodes, and usually combine legal tactics or official discourse (Cai, 2006; Lee, C., 2007). Ho and Edmonds also call them 'depolicitized politics' or 'self-imposed censorship' (Ho & Edmonds, 2008). Regarding the scope of contention, popular protest in China is usually limited to the local level and is rarely directed at the national government (O'Brien & Li, 2006). Even fewer cases reach the transnational level (Lee, C., 2007). O'Brien is correct in noting that protests, especially those in rural areas, are responsive to specific and local aggravation (O'Brien, 2002). Likewise, urban protests are usually based on specific issues and seldom develop beyond the original issue of contention (e.g. Cai, 2006). In addition, Chinese popular contention is characterized as episodic in length. Large-scale actions have occurred that included a series of action forms: collective appeal, petition, peaceful demonstration and spontaneous confrontational actions such as rioting, besieging government compounds, sacking offices and the homes of local bureaucrats. However, actions tend to cease fast once relief is given (Cai, 2006; O'Brien & Li, 2006). Therefore satisfying the demands of popular contention in China is a major goal of resistance.

The organizers or initiators of collective action are an important factor for success (Cai, 2006; Li & O'Brien, 2008; Zhu & Ho, 2008). These individuals are usually local elites or stakeholders with strong charisma. Personal networks are often utilized to mobilize, devise and direct acts of contention. Besides this, several authors have recognized that alliances formed between protesters and the mass media are a distinct movement strategy. Given certain autonomy by the regime, the mass media have tended to act sympathetically towards popular contention (Yang, 2005a; Zhu & Ho, 2008). By exposing the misconducts of local officials, individual citizens have sometimes succeeded in gaining attention from higher-level political officials and in reaching their goals.

When explaining the momentum for public contention, scholars agree that the political system provides space for public contention in at least three aspects. First, it is believed that the limited capacity of the Chinese state to address social and environmental concerns provide opportunities for Chinese NGOs (White *et al.*, 1996). Second, fragmentation exists in the state bureaucracy structure (O'Brien & Li, 2006). The political output system is not working effectively to implement central government's policy. Although the central authorities have put constraints on local officials, these agents' misconduct remains very serious at local levels and affects ordinary people (Bernstein & Lü, 2003; O'Brien & Li, 2004, 2006; Lee, C., 2007). Third, institutional arrangements and legal frameworks have become accessible to the public, through state policy, improved information flows, etc. The public is thus empowered to raise public appeals related to local official misconduct through legal documents and other instruments (e.g. Gallagher, 2007).

Yet such explanations can hardly illuminate the limited visible resistance in contemporary China. Some authors blame the nature of the regime that limits the boundaries for public participation (e.g. Ho & Edmond, 2008). In the field of environmental politics, ENGOs' development and influence are seen as depending on the attitudes and trends set by the state (Brettell, 2000; Ho, 2001). Others argue that the lack of resistance does not always result from government's strong repression. Instead, it is noted that the state bureaucracy's flexibility reduces the number of protests in China (cf. Cai, 2006) by adjusting both its political input and output systems. In the political input system, the complaint system has been revised and institutionalized. Such a mechanism requires that officials are responsible to citizens, and officials must accept citizens' oversight. Bretell's work on environmental complaints in China illustrates how such a complaint system can become an alternative to the legal system for political participation. Its implementation serves as an outlet for environmental grievances and explains the 'loss of momentum toward an environmental movement' (Brettell, 2008). On the other hand, the state apparatus is improving its output system alongside economic reforms. The central government has been active in constraining local misconduct and ensuring effective policy implementation. Such efforts are aimed at maintaining social stability as well as the Party's legitimacy, and have successfully avoided the escalation of social protests (Cai, 2006). One unique feature of the Chinese communist regime among authoritarian regimes is that it sometimes permits, or even encourages, collective citizen action as a way of channeling criticisms and complaints, as long as such actions do not question the central government's status or endanger social stability (Perry, 2001; Cai, 2006; Lee, C., 2007).

Two main themes emerge from the existing body of literature on popular contention in contemporary China. First, focusing solely on state and society opposition may obscure the exact initiatives and outcomes of social protest in China. Future research should avoid exaggerating the effects of individual protests for the sum of all protests, or isolating the role of social forces from state actors (O'Brien & Li, 2006). Second, research on China's popular contentious politics has indicated that in-depth analysis of specific actors—social forces as well as state officials—in given locations is a required strategy. Observation and analysis of movement development and consequences within a certain context are required (*ibid.*). Paying attention to local impacts on the immediate circumstances (McAdam *et al.*, 2005) also applies to the case of China—a country with tremendous differentiation across regions. In this perspective, Chinese scholars are lucky to be able to compare popular contention across different regions in China.

Network approaches in social movement studies

Network analysis is often regarded as a 'paradigm' or a 'perspective', rather than a predictive, normative or substantial 'social theory' (Emirbayer & Goodwin, 1994). Very often, network approaches are used as a kind of structural analysis, a strategy for investigating features of interactions among nodes in a social structure (Wellman & Berkowitz, 1988). It derives from Simmel's formalistic sociology, which focuses exclusively on the overall structure of network ties (Emirbayer & Goodwin, 1994).

As a structural approach, network analysis is capable of providing a general structural analysis on social movement processes (*ibid.*). In this approach, patterns of relations are the basis on which a network theory builds its explanations (Knoke, 1990; Burt, 1992). And both the forms and contents of relations among social positions have significant consequences for the formation of individual attitudes and behavior. This structural component studies individuals' social behavior through their structural relations, rather than the characteristics individuals bear (Wellman, 1983; Wellman & Berkowitz, 1988; Bearman, 1993).

More specifically, in social movement studies, two focuses are often identified in studying networks. Single-mode network analysis emphasizes the properties of networks. It focuses primarily on properties of social ties. It takes social ties as an independent variable and studies their properties (Emirbayer & Goodwin,

1994). Measures of network position include five main principles: cohesion, equivalence, prominence, range, and brokerage (Anheier & Katz, 2005). These principles complement each other in analyzing the features of network structures. They may be used as independent variables (Diani, 2002), and measured to show actors' positions in their networks.[1]

The second focus of network analysis, so-called two-mode network analysis, studies structural linkages in the scope of movement networks. This perspective allows researchers to examine social ties that are found in social events and organizations (Anheier & Katz, 2005). It is suggested that membership in organizations can be treated as affiliation networks and analyzed just like interpersonal networks (Wasserman and Faust, 1994). In this way, a two-mode network analysis is used to map the structural linkages between 'layers' (Anheier & Katz, 2005), when one category (composed of individuals) is tied to the other category (composed of organizations). However, this perspective neglects the case when merely informal interactions—instead of formal interactions among organizations—may happen as a result of personal interactions (e.g. Emirbayer & Goodwin, 1994; McAdam, 2003). So the aggregation of individual networks does not always result in linkages between organizations, to organizational networks (McAdam & Paulsen, 1993; Diani, 2002; Gould, 2003; Passy, 2003).

Aside from its methodological approach, network analysis has also produced interesting outcomes concerning the impact of social ties in movement dynamics. In movement studies, network analysis proceeds in three dimensions that are emphasized by classical social movement theories. First, this approach focuses on the process of resource mobilization in movement dynamics. Second, the formation of cooperative actions is studied through networks. Third, network approaches are powerful in explaining the establishment of collective identities in movement processes. I will now elaborate on these three dimensions.

First, in the literature discussing individual interests and collective goods, social networks and connections are argued to be instrumental in mobilizing resources, such as the money, power, reputation, or manpower of social actors, which can be gained through social relations (Burt, 1992; Lin, 2001). Participation in movements is of particular interest. Oberschall (1973) refers to cells, branches or groups of members as a mechanism of 'bloc recruitment' with mobilizing effects. Individuals with pre-existing ties are more likely to be contacted and recruited to movements than are individuals without such ties (Snow *et al.*, 1980; Gould, 1991). Putnam explains citizen participation and civic engagement in terms of 'social capital', which refers to connections among individuals—social networks and the norms of reciprocity and trustworthiness that arise from them (Putnam, 1993). He argued that social capital allows community members to overcome the dilemmas of collective action, which would otherwise hamper their attempts to cooperate for the purpose of improving social life. Various scholars have also shown that the number of ties movement activists have is correlated with the number of individual participants (Rosenthal *et al.*, 1985; McAdam, 1986; Gould, 1991; Friedman & McAdam,

1992). As Tarrow summarizes, social networks have emerged as the most common sources of recruiting new members into social movements (Tarrow, 1998: 124).

More specifically, social networks have differential levels of affecting mobilization, because of their linkages to their social contexts. Ohlemacher (1996) demonstrated that closed networks in which protesters were relatively isolated from their community failed to generate broad mobilization, while protesters who had substantial non-protest ties with their community built a broader, less marginalized mobilization. At the same time, local conditions affect the function of a social network. McAdam and Fernandez found that successful recruitment to the Freedom Summer campaign depended on the location of organizational networks. Campuses with a weak tradition of activism are more likely to encourage individual participation than campuses that have a strong tradition of alternative politics (McAdam & Fernandez, 1990). The more a movement is different from the dominant orientations in society, the more effectively certain networks seem to function (Della Porta & Diani, 2006).

Second, social networks play a crucial role in the organization of a social movement (Gould, 1991; Schlosberg, 1999). Not being constrained by boundaries that distinguish different social organizations (Diani, 2003), personal networks may connect actors in different parts of the movement (Diani, 1995; Tarrow, 1998). Research has shown how pre-existing networks provide a social resource in establishing coalitions. Hurlbert *et al.* (2000) detailed how individuals used their core network of strong and dense ties in helping following a hurricane in Louisiana. Even SMOs that are, in many ways, peripheral to the mainstream of society accomplish their goals by using networks (Jenkins & Perrow, 1977). As the internet has begun to play an important role in organizing collective actions, Pickerill examined three British environmental groups' computer-based communication methods and their implications. Through email, electronic newsletters and bulletin board systems, computer-based communications enable individuals to have a larger scale of freedom when expressing their ideas. They also provide efficiency for SMOs in organizing activities, for instance, faster exchange of information and networking ability, and increased diffusion of ideas and tactics across social movements and geographical areas. At the same time, the hierarchy of organizations can be reduced significantly (Pickerill, 2001). Despite the fact that computer-based communication operates at its best when backed by real social connections in specific communities (Hampton & Wellman, 2001; Tranvik, 2004), it is reasonable to expect the internet to be increasingly influential in the organization of social movements by connecting all sorts of communities that do not have apparent links (Della Porta and Diani, 2006).

Third, the functions of networks are not only limited to the micro-level in affecting motivations of individual activists, but at the macro-level they facilitate the formation of collective identities. A sense of collective identity is very important to social movements (McCarthy & Zald, 1988; Klandermans, 1989). Networks can be quite instrumental in the formation of public consciousness, and in further facilitating collective identity. On closer examination, it is obvious that to mobilize individuals to participate in collective actions is not simply to build links among them, rather it is a dynamic process of meaning construction. Existing as a particular focus that represents their social contexts (Guthrie, 1998; McAdam & Paulsen, 1993), social networks facilitate identity construction and identity strengthening (Passy, 2003). Broadbent's case study of Japanese social networks demonstrates that Japanese networks act as vehicle of cultural meanings and interact strongly with cultural factors (Broadbent, 2003).

Among all types of social network, one special type that has been the focus of considerable scholarly interest is elite networks. One example of this kind of network is composed of ties between politicians and corporations when a single individual sits simultaneously on the boards of directors of both organizations. Elite networks not only facilitate communication among elites, but are also influential for social groups to access political power and influence policy agendas.

Individuals in elite networks may have increased access to political power. Useem's work examining directors of large corporations in the USA revealed that inner group members—consisting of a small number of the elite who simultaneously hold multiple directorships in major corporations—are more likely than single directors to serve on federal advisory boards and on state and local government boards (Useem, 1979). Moreover, the structural positions in elite networks are important for predicting individuals' influence in policy issues. Knoke conceptualized these networks as a number of concentric circles that indicate power inequalities according to actors' network positions (Knoke, 1993).

Network relations in the political arena can have effects on event outcomes and influence policy agendas. According to Coleman's exchange model, actors are more powerful and more likely to achieve the outcomes they prefer if they possess resources that other actors have strong interests in (Coleman, 1973, 1977). By examining network relations in energy and health domains (Laumann *et al.*, 1985), it is demonstrated that the more essential one actor's resources are for another, the greater the extent of dependency in the exchange relation. By creating resource-exchange dependency, some organizations succeed in gaining control over policy events of interest to them.

Despite the growing importance of the social network perspective in social movement research, this approach has its own limitations and biases. Aiming to bridge structural lines of analysis and player- or agency-focused studies, such perspectives may lead to the problem of over-exaggerating the relational function of individuals and neglecting other features (Emirbayer & Goodwin, 1994). Methodologically, the roles of different levels of social networks are not clearly distinguished. Organizational networks are established by joint activities, joint staff and joint strategies of organizations, and are assumed to be an aggregation of individual networks (Curtis & Zurcher, 1973). The network approach is also criticized for being too static. Gale pointed out that alliances within and between organizations can shift; some may support actors on one issue and oppose them on others. So some actors may mount counter-movements to social movements (Gale, 1986). As Brass & Burkhardt (1993: 444) described, network ties are stable patterns that 'represent a constraint on behavior'. In policy network analysis, this approach may also 'oversimplify a very complex process and seems to assume that a particular organizational unit has only one set of values and preferences'. In addition, analysis of elite networks' influence is also argued to make the error of equating a high potential for control with actual power (Dahl, 1958).

Environmental NGOs and networks: towards a conceptual framework

In analyzing the development of NGOs in China, I develop in this section a classification of movement organizations along two dimensions. The first dimension focuses on the types of social network that are primarily utilized and adopted in movement dynamics, and follows our discussion above. The second dimension relates to the degree or level of formality of movement organizations.

Personal and organizational networks

Generally speaking, social networks refer to the collection of relations and ties. In social movement studies, social networks are not often distinguished further. As indicated above, however, social networks within social movements should be distinguished in organizational and individual networks. Too often, personal networks composed of personal ties are mixed with organizational networks, where organizations are related to each other.

In addition to studying the organizational development of SMOs, the capacity and actions of individual activists in social organizations also require attention. When key and leading activists play a significant role in promoting and developing the organization, their position, role and identity in social groups may differ. It is often believed that in formal, well established, institutionalized and professionalized social groups, individual members are 'defined' by the groups they belong to. As Simmel (1955) has suggested, group membership strongly affects the identities of individuals. Each person is associated with the set of organizations to which he or she belongs, strongly shaping their identity. So the functioning of individuals in formal social groups is, to a major extent, structured by these groups. At the same time, individuals cannot be fully reduced to the organizations they belong to, but remain semi-autonomous actors who can make a difference. These personal connections and networks of group members can be institutionalized as organizational actions and resources, and utilized as such. Subsequently, these personal attributes contribute to structuring the organizational profile. However, in young, informal, immature organizations, the identity of individuals is less associated with and structured by the organization's profile, and instead continues to remain more autonomous. Personal ties that are mobilized at the individual level are not likely to be automatically institutionalized and remain strongly personal (and thus disappear with the departure of the individual from that organization). Individual activists may preserve their own personal identities, motives, and strategies, rather than being fully shaped by their organization. Individual actors, and their relations, cannot be reduced to or equated with the organization they belong to, but are themselves semi-autonomous actors. Hence we need to distinguish conceptually the two types of network in social movements.

Personal networks consist of a collection of social ties or relations a person may have with other individuals. Personal networks rely mostly on informal personal ties and relations, such as friends, relatives, neighbors, etc. When these personal networks are utilized in the context of SMOs, they primarily carry the profile and identity of the person, instead of the characteristics and identity of the organization the individual belongs to. In personal networks, the positions of individuals cannot readily be replaced by other individuals belonging to the same organization.

Organizational networks refer to a network of links between organizations, which are based on organizational identity and organizational interests, and are usually institutionally embedded and organized. In contrast to personal networks, in organizational networks the persons involved can be replaced by other staff members within the same organization without significantly changing the nature and functioning of the network.

It goes without saying that personal and organizational networks are an ideal–typical construction. In the reality of social movements, we will often find diverse mixtures of individual and organizational networks. Nevertheless, these ideal–typical networks give us an indication of the level at which movement networks are more or less characterized and organized as personal or organizational.

Formalization of environmental NGOs

According to Staggenborg, who works within the tradition of resource mobilization theory, the formality and institutional levels of SMOs determine how they perform within the social movement. Their internal organization and structure is directly related to the strategic and tactical maneuvers of SMOs (Staggenborg, 1988). Thus SMOs with a formalized structure tend to adopt more institutionalized tactics and work through organizational networks, while those with less formalized structures will be more likely to rely on individual and personal networks. Furthermore, it is believed that informal SMOs have a tendency to develop a formal organizational form in the maturation process of the movement (cf McCarthy *et al.*, 1973, 1977; Gamson, 1975: 91). In the development of the social movement, informal SMOs are more inclined to become institutionalized entities (Cable, 1984). Besides this, informal SMOs also differ from formal ones in the different roles they play in movement development. Although informal SMOs are predicted to transform into SMOs, they remain an important protest vehicle in the initiating stage of movement (Staggenborg, 1988), and often still have a (smaller) place in more mature movements

While these conclusions are based on research on social movements in Western democracies, they do not necessarily hold for China. Therefore it is necessary to distinguish the level of formality of SMOs from the networks that are applied.

Three aspects characterize the levels of formalization and institutionalization of SMOs: the membership system, the internal institutions, and the mechanism of decision-making. With these characteristics, SMOs can be placed on a continuum, ranging between formal and informal organizations (Table 2.1).

First, the membership system relates to the degree of collective identity among those associated with an SMO. In informal SMOs, the constituency is often not clearly specified, which is indicated by a poorly established membership system. Procedures of recruitment and maintenance of adherents are not institutionalized. The distinction between inside members and outsiders is often unclear, flexible, and dynamic. Actors who initially do not belong to a SMO may easily become involved in the group's activities, and even assume important positions, in a relatively short time. The boundary between this kind of SMO and its social environment is fluid. All this will result in a limited degree of adherence between members and a low level of shared identity. In contrast, a well established membership system will boost a sense of belonging among members and increase the common identity of the organization.

Second, the internal institutional system—the rules and norms that govern the internal organization—is not properly established and articulated in informal social organizations. A social organization is a system that coordinates people's behavior by means of rules and norms (Katz & Kahn, 1966). And rules and norms direct the attitudes and behavior of an individual belonging to that organization. Without the formal establishment and articulation of these institutional rules, an SMO remains an entity that is bound by voluntary, personal relationships. In analyzing the extent of formality in the internal organization, the relationship between leaders and members is distinctive. In informal groups, leaders of environmental groups usually have personal charisma, through which they play a strong role in mobilizing constituency and directing the organization (Gottlieb, 1993). Relationships between

Table 2.1 Formal and informal organizations

	Membership system	*Institutional system*	*Mechanism of decision-making*
Formal organization	Well established	Clear and articulated rules and norms	Well established
Informal organization	Poorly established or does not exist	Poorly established or does not exist	Absent or poorly formalized

leaders and members are not established by formalized rules and institutions, whereas in formal organizations such interactions are characterized by written regulations and articulated institutions. The relationship between members and leaders will be clearly defined, and job descriptions usually establish the expectations of both. Internal differentiation usually is established by the processes of internal articulation and formalization.

Thirdly, in informal SMOs a mechanism for internal decision-making is often absent, or poorly defined and formalized. Consequently, individual members will have little power in deciding organizational profiles, priorities, and strategies. In formal social movements, these internal decision-making processes are clearly structured and formalized, providing clear points of access for members in decision-making processes.

Classifying environmental NGOs

Regarding the nature of SMOs, a two-dimensional classification can now be established (Table 2.2). On the vertical axis, the formality of environmental NGOs is indicated, with two types of organization: informal ENGO characterized by informal internal interactions, and formal ENGOs that have more institutionalized internal structures. On the horizontal axis, the types of social network that prevail in the strategies of the ENGOs are indicated. This results in four ideal–typical ENGOs.

This classification enables us to hypothesize and investigate the utilization of social networks from a dynamic perspective, and also related to their specific contexts. In the evolution of the environmental movement in Western societies, movement organizations of different levels of formality differ in their adoption of social networks. Personal networks are usually found in relatively young and immature SMOs. Environmental movement organizations of high formality generally use institutionalized and formalized internal structures to reach their goals, and work primarily via organizational networks. As to the evolution of ENGOs, the environmental movements in Western countries have demonstrated that type IV SMOs have become the dominant type in the contemporary environmental movement, with high levels of professionalization.

Table 2.2 Types of ENGO in environmental movement

Formality of ENGOs: level of institutionalization	*Principal networks*	
	Personal	*Organizational*
Low	I (informal organization, utilizing personal networks)	II (informal organization, utilizing organizational networks)
High	III (formal organization, utilizing personal networks)	IV (formal organization, utilizing organizational networks)

The type IV ENGO is the most influential type of organization among Western societies, particularly at national level. Although the other three types of ENGO still exist in Western Europe, they usually prevail at the local or grassroots level, and arguably their social and political impact is less prominent. The central question in this study is whether that is, or will be, the same in China, which has a context that differs significantly from Europe and the USA.

Networks and Chinese environmental NGOs

This section analyses the utilization of social networks in the Chinese environmental movement. As personal networks constitute a cultural phenomenon in Chinese society, this section starts by elaborating on the specific cultural and institutional backgrounds of such networks in China. Subsequently, the role of personal networks in the contemporary environmental movement in China is elaborated.

Social networks in China

It is not accidentally that Castells (1996, 1997), in his major triology on the network society, pays ample attention to China. Chinese society is a network society, as it is full of complex personal and often informal connections, or *guanxi*, among friends, relatives, colleagues, neighbors, and so forth. *Guanxi*, loosely translated as 'connections', is a specific Chinese idiom for characterizing the social networks in China. The functioning of *guanxi* is integrally linked to other building blocks of Chinese society, such as *ganqing* (sentiment), *renqing* (human feelings), and *mianzi* (face) (cf. Gold *et al.*, 2002). In this study, *guanxi* refers to a personal network. And the term *guanxi* network will be used as an alternative to personal network.

The literature on the importance of Chinese interpersonal ties and connections (*guanxi*) in the functioning of individuals and the structuring of wider social relations and systems is not recent (Gold, 1985, Walder, 1986; Lin & Bian, 1991; Yang, M., 1994). In explaining the strong position and formative power of *guanxi* in Chinese societies, most scholars refer to the specific Chinese cultural origins and the specific institutional context. I subsequently explain these cultural (this section) and institutional (next section) contexts.

Guanxi and personal relationships are regarded as an indigenous cultural phenomenon that prevails throughout Chinese society (Hsu, 1985a; Hwang, 1987; Yang, M., 1994; Hwang, 2000; Chen & Chen, 2004). What is more, *guanxi* studies focus closely on the construction process of personal networks and the exchange of resources and favors through these personal networks. To these authors, personal relationships are rather independent and often understood to be separated from group associations. So the fundamental unit of *guanxi* is the individual, rather than the group in which these individual persons are located (Alston, 1989; Yang, M., 1994). This aspect distinguishes *guanxi* studies from a large number of substantial studies on interpersonal relationships (e.g. Rubin, 1973; Berscheid, 1985; Berscheid *et al.*, 1989) and social networks (Coleman, 1973; Granovetter, 1973; Carley, 1991; Burt, 1992) that focus primarily on the structure of networks and the positions of individuals and organizations in such networks (Brass, 1995; Chen & Chen, 2004).

Chinese cultural values are believed to have strong implications for the individual behavior of utilizing personal ties and connections. Confucianism, in particular, is believed to constitute the historical and cultural roots for the importance and widespread functioning of personal relationships as structuring elements of social development (King, 1991; Chen & Chen, 2004). Three criteria of Confucianism (*ren*, *yi* and *li*[2]) are understood to direct people's behavior in their social networks (Yum, 1988). At the same time, one's relationship is very dependent on others, instead of taking the world as an opponent to oneself (Hsu, 1985a). According to the directive of Confucianism, Chinese social relationships appear as an intrinsic phenomenon. Chinese individuals view and identify themselves through their personal and social relationships, and behave according to, and rely strongly on, the position they have in their relational context (e.g. Hsu, 1985a; Hwang, 2000). This aspect differs greatly from individuals in Western society. While Western individuals equally have personal connections and ties, these are less influential on and decisive for their behavior and the structuring of social systems. Directed by individualism, Fei sees Western individuals as akin to wooden sticks, which may be bound together by their social organization just like a bundle of sticks (Fei, 1947). In contrast, he describes individuals in Chinese society differently (*ibid.*):

> 'The structure of Chinese society is like ripples caused by throwing a stone into a pond. Everybody is situated at the center of the water rings ... No matter when and where one finds oneself, one is always situated at the center of the flexible social network.'

Two perspectives can explain this difference. First, Chinese society brings forth personalities that differ from westerners. The Chinese or even Asian personality represents a *Galiean* view of human nature, also called as 'authoritarian personality'. Westerners have a *Ptolemian* view of human nature, also called 'democratic personality', which is deeply rooted in a Western culture of individualism, and regards individuals as situated at center of the universe (Inkeles, 1972; Hsu, 1985b). Second, this feature of building connectedness in Chinese society is derived from kinship connections (Yang, M., 1994), which are rooted in family relationships as a primary type of relationship in individual interactions in China (Goode, 1959;Yin, H., 2002). As inferred by Metzger (2001), Chinese familism not only was basic to traditional society, but also has been vital to all the successful modern Chinese societies.

Social networks and the political system

The significance of personal networks is secondly understood as a phenomenon that is placed in, and relates to, the specific institutional context of China

(Gold, 1985; Walder, 1986; Yang, 1994; Bian, 1997; Guthrie, 1999; Wank, 1999, 2003). *Guanxi* complements and supplements bureaucratic and formal legal interactions and structures that are so present in contemporary—but also in past—Chinese society (Wellman *et al.*, 2002). In the hierarchical one-party political system, individual networks are an essential element in successfully accessing governmental authorities, while at the same time they preserve the strength of the hierarchy and elitism of the political system (Guthrie, 1998; Xia, M. 2000). In addition, *guanxi* proves to be functional in the economic realm as well. In China, 'private' economic actors tend to rely on and make use of *guanxi* to organize the necessary resources to keep their business going, in addition to, and sometimes even in place of, market economic incentives and rules (Guthrie, 1998).

Under the influence of Chinese political culture, individuals tend to rely on informal mechanisms rather than formal structures for their political participation. Generally speaking, Chinese citizens and social movements have cautious political behavior, and will not participate in opposing politics, as their Western counterparts tend to do. The monopoly role that the Party state has over society is widely acknowledged and hardly ever contested. Pye, a scholar studying the cultural dimensions of Asian power and politics, remarks that 'in the Chinese scheme of things all legitimate power was [traditionally] limited to officialdom, and no significant forms of power were supposed to be at the command of any other element of the population, regardless of social station' (Pye & Pye, 1985). So the formality of their organized actions is relatively low. As introduced in Chapter 1, the political authorities have kept strong control over the development of civil groups. Voluntary organizations have thus encountered difficulties in their organizational growth and institutionalisation. As demonstrated by an empirical study on political participation behavior in Beijing, citizens are inclined to use informal and individually based methods, which further indicates that the Chinese prefer to interact with individual political elites rather than to follow other types of interest articulation (Shi, 1997).

What is more, the behavior of Chinese political leaders also, to some extent, leads to low institutionalization of social organizations. Chinese political elites play an important role in political processes. However, influenced by Chinese philosophies, Chinese political leaders are very pragmatic and flexible, can quickly adapt to different situations, and may not maintain stable political ideas in their posts (Pye, 1986). Their informal interactions with the public can often affect the decisions and policies they make (Shi, 1997). In this way, the public can affect political leaders without being formality organized.

Thus, due to the authoritarian political institutions in China, informal practices are extensive in Chinese political processes. This forms a unique feature of Chinese political culture, and constitutes an important factor in determining the adoption of personal network strategies in the environmental movement.

Functions of social networks in the Chinese environmental movement

In the environmental movement, personal networks function primarily in three specific ways. First, with the characteristics of Chinese political institutions, personal ties are used in accessing the political system. Second, personal networks facilitate the formation of collective identity. Finally, personal networks promote the establishment of coordinated actions.

Accessing China's political system

Interactions with the political system constitute one important aspect of using personal networks in China's environmental movement. As governments at different levels are merely monopoly actors in environmental governance, environmental activists have to interact with the Chinese political authorities if they want to influence environmental governance. Their interactions with governmental and political authorities focus on three aspects: accessing environmental information, acquiring political protection, and influencing policy processes. Their ultimate goal is to participate in policy processes and generate influence at this stage.

First, using social networks is important in obtaining useful information on environmental governance. While it is in a process of transition (cf. Mol & Carter, 2006), the Chinese bureaucracy is still hierarchical, closed and non-transparent, and only limited formal mechanisms and structures exist for acquiring information that is not directly freely available. Unlike their Western counterparts, who mainly rely on institutionalized and formalized procedures and structures to access information, Chinese SMOs rely strongly on the practice of *guanxi*. As argued and illustrated by Bian (2002), *guanxi* will work as effective channels of information when information diffusion is either restricted or ineffective.

Second, personal networks are used to acquire political protection and recognition from the current political system. As established, the Chinese citizens have a tendency to regard the government as the authority for public administration, and usually refrain from opposing it. Under this situation, ENGOs, as representatives of the public, have to acquire a certain level of protection and recognition from government agencies before they can produce social leverage. As mentioned, the legal framework on NGOs in China is weak at present, an additional reason for relying on government protection rather than on legal protection. Informal interactions and personal networks play an important role in gaining and safeguarding such protection.

Third, personal networks are used to generate power and influence and to participate in policy-making processes. Environmental issues are not a major priority in the Chinese government. Via individual networks with key officials in the government, environmental NGOs have a better chance of influencing the political agenda, and are acknowledged by government as participants in the discussion of policy processes. In this way, articulation of interest in the political arena is possible without access to formalized channels. This becomes even more relevant when the Chinese state is losing its monopoly position in (environmental) governance.

Facilitating the formation of collective identity

Research has illustrated how social networks facilitate the establishment of common identities. Social networks are a mechanism for spreading information and conducting effective communication. Through personal networks, people get to know others with similar interests, and they tend to validate information by comparing and discussing their interpretations with other members of these networks. Established networks also allow leaders to mobilize large groups of actors without convincing each member personally of the virtues of the cause (Klandermans, 1984; Marwell *et al.*, 1988; Friedman and McAdam, 1992).

In China, *guanxi* networks also constitute a trust-building and -maintaining mechanism, which is formative in mobilization (Bian, 2002). It is suggested that the development of *guanxi* is based on commonality, shared identity, and trust (Jacobs, 1979). Each person in a *guanxi* relationship shares an aspect of personal identification that is important to him or her as an individual. This shared identification is a necessity in *guanxi*. When solidarities are constructed according to the Chinese cultural and ethical system, they form close connections. So key organizers play important roles not only in linking people together, but also in mobilizing people and forming and framing the collective identity of environmentalism. The formation of environmentalism as collective identity seems especially relevant for young people in contemporary China, as they do not have any particular and definite goals, framings, and commitments. As a result, young people comprise a significant share of the members in the environmental movement. In addition, values of collectiveness, universalism, and benevolence, which are prevalent in traditional Chinese culture but are also building blocks for Chinese *guanxi*, parallel well with pro-environmental values and behavior.

Promoting coordinated actions

In China, the context for environmental NGOs and their activities is not encouraging. ENGOs are competing over limited resources and information flows, and cannot always rely on public visibility and open coalition-building. Personal networks can be important in promoting collaboration between environmental NGOs and a variety of other actors.

Experiences in many developing countries have shown that networks play a crucial role in strengthening the capacity of ENGOs when the latter are weak in resources and capabilities. Networks substantially facilitate communication with other constituencies inside and outside the environmental movement, within a locality and beyond that (Royo, 2000). The loosely structured forms of networks can link both large and small organizations efficiently, and help local NGOs to grow and develop.

In addition, Chinese personal networks allow interest exchange between private and public affairs (Hwang, 1987). In Western societies, contracts, laws, and institutions govern conflicts between private and public interests (Yum, 1988). In China, with many of these institutions, contracts, and laws still in the making, individuals and social movements use the practice of *guanxi* and personal connections to mix private with public affairs (Wang, Y., 1987; Yum, 1988).

Methodology

Advantages of case study research

Here case study research is applied as the main methodology to examine the impact and functioning of personal networks in the environmental movement in China. Case study methodology is used for two reasons.

First, according to Yin, case study research is especially relevant if 'how' and 'why' questions are asked about a contemporary set of events, over which the investigators have little or no control (Yin, R., 1984). Case study research can describe and explain complex and entangled group attributes, patterns, structures, or processes (Verschuren, 2003), and thus is suitable to the study of the Chinese environmental movement, which is characterized by interactions and dynamics that vary largely in specific locations.

Second, to examine the roles and functions of personal networks, we require case studies. Case study research forms a suitable strategy to study social networks in their specific context, because it can generate prolific, detailed elaborations and understandings of variations in social phenomena (Snow & Trom, 2002). So case study research, or qualitative field research, may produce strong evidence in exploring the relationship between network composition and the position of organizations in macro-level networks (McAdam, 2003).

However, while providing in-depth understanding of particular cases, this methodology also has its weaknesses. By utilizing case study as a research methodology, the research has ambiguities in both its internal and external validity (Verschuren, 2003). Externally, this study will be less powerful than quantitative methods to provide generalized theoretical propositions. What can be learned from this study will primarily be regarding the specific environmental movements in four case-study cities, rather than representing the universal features of the whole country. Yet, through these four cases, this study represents a variety of local conditions in political circumstances, economic levels and geographical locations. So, by utilizing multiple cases, this study has the potential to contribute to theoretical refinement of environmental movement theories (Snow & Trom, 2002), especially with the addition of the conclusions of other studies on environmental movements in China.

This study also has limitations with regard to internal validity, which is researcher-dependent (Creswell, 1994; Verschuren, 2003). This weakness is unavoidable, regardless of the techniques utilized in data collection. Particularly, as the focus of this study is a phenomenon that is related to a variety of conditions within a local context, the use of participatory observation and unstructured, open-ended interviews will probably add to problems of internal validity, but at the same time provide abundant data and insight.

Selection of case studies

This study is made up of case studies of environmental activism in four cities. Each case starts with an overview of basic information on the city, including its local culture and political institutions. Subsequently, each case investigates the environmental networks comprising environmental NGOs and their communities. From these networks, sub-cases are selected, consisting of an environmental NGO or a personal network of activists. I introduce the selection of cities as sites, before specifying the selection of sub-cases.

Selecting sites

In this study, Beijing, Shanghai, Panjin (in Liaoning Province) and Xiangfan (in Hubei Province) are selected as cases to examine the role of personal networks in environmental movements (Map 2.1). Three criteria were applied in

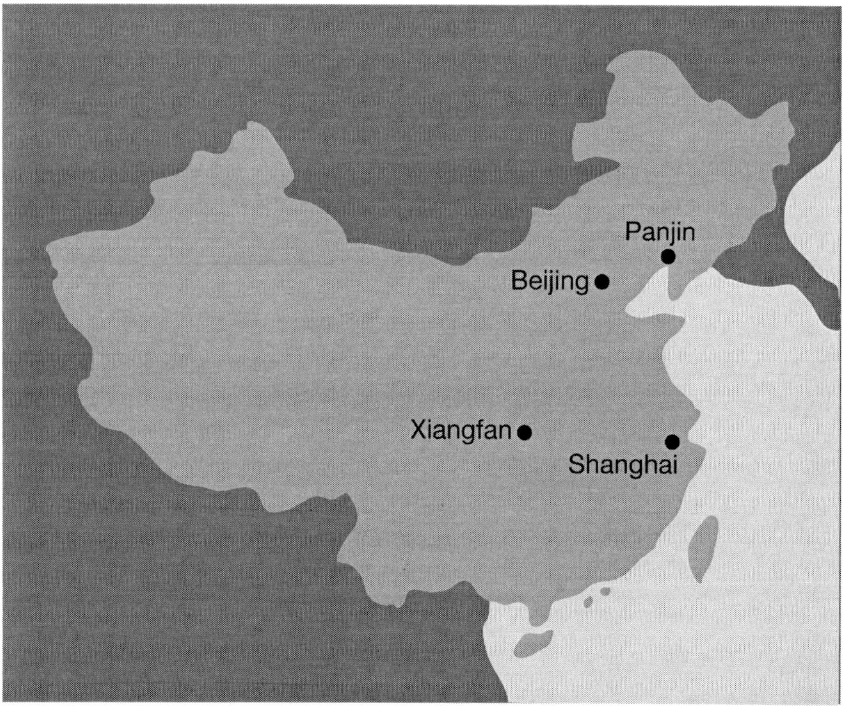

Map 2.1 The four case-study sites

selecting these cases: economic development, openness of the political system, and local culture. From the theoretical review above, it can be expected that these dimensions may affect the practice of personal networks in the environmental movement (Table 2.3).

First, the level of economic development is considered in selecting cases, in terms of GDP per capita. It has been established that economic development may affect environmental activism in two ways. High economic development exerts pressure on the environment (e.g. Lee & So, 1999). And several scholars have indicated that high levels of economic growth go together with a middle class that is often a vehicle for environmental awareness, and with more resources for building environmental capacity. Both may affect both the degree of environmental activism and the organizational modes.

The second criterion used for selecting cases is the openness of the local political system towards civil society organizations. A repressive local government probably will not facilitate a prosperous development of formalized environmental NGOs, and will also affect the movements' action repertoire and strategies. Variations in the political system may provide us with variations in the use of personal networks.

Third, local culture is used as a criterion for selecting cases. As indicated above, the adoption of personal networks in Chinese society is often seen as an essential part of 'China's national character' (Hwang, 1987). Yet cultures also differ within China, for instance between North and South, and between the more cosmopolitan coastal regions and the inland provinces. The use of personal networks in the environmental movement is more likely to prevail in regions with a strong traditional Chinese philosophy and a strong influence of religions, particularly Confucianism (e.g. Jacobs, 1979; King, 1991; Ji, 2006). So in regions that have essentially diverted from ancient history and traditional cultural influences, the use of personal networks might differ.

These criteria are not used for formulating and testing hypotheses on the relationships between these contextual factors and the prevalence of personal networks. That would require a different kind of research design and much larger research effort. The criteria are used mainly to select cities that, together, capture sufficiently the diversity of local contexts against which environmental activism emerges in China. This research is more exploratory in character, and focuses more on the question of how local contexts influence the functioning

Table 2.3 Criteria for comparison of cases

	Economic development level	*Openness of political system*	*Local culture*
Shanghai	High	Very repressive	Cosmopolitan
Beijing	High	Rather open	Mixed
Panjin	Low	Repressive	'Traditional'
Xiangfan	Very low	Repressive	Mixed

of the environmental movement in urban China, with special reference to personal networks.

Selection of ENGOs

In each of the cities, at least one environmental NGO is studied as a sub-case study. And one environmental campaign involving that NGO is focused upon. Selecting ENGOs as sub-cases was simple in two sites (Xiangfan and Panjin), as the size of the environmental NGO community was small and only a limited number of registered environmental NGOs existed. The reason that few registered environmental groups exist is also related to the Regulations on Social Organizations, promulgated in 1989. According to this law, only one organization of any one type is allowed to register at each administrative level.

In the other two cities, the focus was on selecting the more successful NGOs. Three aspects were used as criteria. First, the success of environmental NGOs can be judged by the annual budget. The shortage of funding poses a serious problem to most grassroots NGOs in China (Schwartz, 2004; Wang, A., 2005), so the environmental NGOs with larger quantities of financial resources prove to be successful. According to a survey, the majority of ENGOs manage an operating budget of less than 100,000 RMB (approximately US$15,000) (Deng, 2001). The ENGOs chosen in these two cases have a budget of 100,000 RMB or more per year. Second, ENGOs with substantial social leverage can be assessed by the number of participants (total number of volunteers and staff). Third, the environmental NGOs to be included needed to be established before the year 2003, so that they had time to build their network.

Methods of data collection

The empirical research was conducted between December 2003 and August 2005, during which time the author was stationed in Beijing, as a member of the research staff in the NGO Research Center in Tsinghua University. This center investigates and studies the development of Chinese NGOs, and is well known among the NGO community around the country. Identified as a researcher for NGOs, the author gained access to environmental NGOs and was accepted to conduct participatory field work. This enabled the use of participatory methods as the principal means of gaining information. The author participated in several projects and numerous gatherings related to campaign building of various NGOs.

Triangulation of methods is also used (Verschuren & Doorewaard, 1999). Direct observation, in-depth interviews, and archival documents have been gathered to garner information. In-depth interviews have been held not only with key figures in environmental NGOs, but also with more peripheral members of the organizations, with governmental officials, Environmental Agency officials, the opponents of environmental activists, and other relevant actors (see Appendix). Documentary information collected includes records of meetings, biographical information, personal emails/papers, archives of NGOs, and address books. In addition, published sources such as newspaper articles in national and local newspapers and news reports were systematically collected during and after campaigns.

Collecting information on networks

Information on networks forms an important part of this study. This includes mainly data on personal relations, interactions and networks, and the operation of environmental groups. In addition, information on the ongoing movement and coordinated actions is also collected for this study.

With regard to personal networks, identifying the scope and composition of a network are two important prerequisites to analyzing its function in the dynamics of social movements. The boundary problem is important in the data-collecting process. In this study, informal networks are not strictly demarcated: all individuals who interact with the focal persons in matters related to environmental activism are included. The method of snowball sampling is used, through which the individuals in a personal network are selected and introduced one by one, limiting the chance of omitting some key figures. Basic information about the individuals as nodes in the focal person's personal network is collected in order to map network characteristics. For example, general background information about individuals is collected, such as occupation, age, education, and income level.

Data on organizational networks is collected primarily by investigating the profiles of ENGOs and their connections and collaboration with other organizations, inside and outside the environmental movement. Written material such as pamphlets, newsletters, websites, and working journals provide a primary source for organizational links. In addition, news coverage by the media provides a second source. But the chief source for organizational network data comes from interviews with staff of ENGOs, with government officials, and with NGO-watchers such as academics and foreign experts.

3 Environmental activism in Panjin, Liaoning Province

Panjin's socio-economic context

Panjin, a city of 1.2 million people, is located in the triangle of Liaohe in the Liaoning Province. Panjin used to be dedicated to growing grain under the centrally planned economy in the 1950s. Along with the discovery of oil fields in Panjin in the 1960s, the county speeded up its industrialization process. Only since 1984 has it been established as a city whose administration governs two counties and two districts.[1]

During the past 20 years, the underdeveloped county has experienced fast economic growth, and has been transformed into a modern city. More than 80 billion RMB has been invested in this process to build roads, water supply, heat supply and other city infrastructure. The city's environment has also been changed notably. In 2003, lawns and plants covered 25 per cent of the city; a proportion very close to the national average of 27 per cent. Panjin's economic development has been fast in the past 20 years. The average income of its urban population is approximately 6050 RMB (approximately US $746.0), and 3470 RMB (US$427.9) in rural areas. It has been listed as one of the 36 cities with the highest standards of living around the nation.[2] When examining the economic increase in the city, it is not difficult to discover that in the past years Panjin has largely been dependent on the Liaohe oilfield, the third biggest oil reserve in China. More than 50 per cent of the total economic income in Panjin was achieved from oil and gas in past years. Due to intensive drilling, resources are now almost depleted, and economic development is slowing down. Recently, the city government began to promote the development of ecological agriculture and industries that are related to petroleum.[3]

The urban environment of Panjin city was realized after its economic development. Seen from its economic structure, the main economic growth—promoted by the oil industry—was dominated by the state-owned economy. The private economy started only in the mid-1990s, and developed relatively slow.[4] Until 2005, the private economy accounted for less than 25 per cent of the GDP. This was still seven percentage points below the provincial average, and 25 percentage points lower than the national average. Furthermore, Panjin's citizens have a low education level. Only one university has been built in the city, and 5.8 per cent of the population have a university education, 6 per cent fewer than the average in Liaoning Province in 2002.[5] As a consequence, activism in developing grassroots environmental groups is very weak: there are only one or two grassroots environmental groups in Panjin.

The government has still complete responsibility and authority in governance. Because of the strong dominance of state-owned enterprises in the economy, authorities play a major role in local political and economic development. But the local political atmosphere is relatively conservative towards allowing NGOs into the policy process. Local political authorities have very poor knowledge of environmental NGOs (ENGOs) and almost no interest in giving them positions. In this situation, little space is available for the latter to be accepted by political institutions.

Aside from the political authorities' attitude toward NGOs, citizens show weak activism in establishing grassroots NGOs. In recent years, individual citizens have started to become more concerned with environmental protection, using hotline calls, mail and visits to report environmental misdeeds. In 2003, around 800 hotline calls, mails or visits were directed to local environmental protection agencies to report on environmental misdeeds. The cases most often reported include air pollution, noise pollution, and solid waste management in the city.[6] This number is larger than the average number of calls across the country.[7] But the environmental concern of Panjin citizens remains at a relatively low level and hardly develops into articulation through organizations. To local citizens, grassroots groups are of little importance in environmental governance. As state agencies are regarded as the main units for environmental governance in Panjin, civic activism and the development of environmental voluntary groups is limited.

The inclination of Panjin's people toward the development of civic organizations may lie partly in local culture. Located in Dongbei or the Northeastern Region of China (covering Heilongjiang, Jilin, and Liaoning provinces), the majority of Panjin's inhabitants originally came from all three provinces (Xu & Sun, 1996), and bear the cultural heritage of this area. As one of the largest regions in China, Dongbei has a multi-ethnic heritage. It was populated by the Xianbei (a nomadic, proto-Mongol people) during the Han dynasty (25–200 AD), then the Qidan (semi-nomadic Mongolian people) during the Tang and Song dynasties (618–1279), and the Nv Zhen (also called Jurchen, a Manchurian people) from the Song to the Yuan dynasties (960–1368). The frequent shifts in governance left this region with little wealth and capital compared with the South of China. It was only during the Qing Dynasty (1644–1922) that Dongbei was administered by a relatively stable administration of the Manchu minority. Government instability fostered a culture of migration. In comparison with the southern Chinese cities, where many of the ancient Chinese dynasties were located, the cultural achievements in Dongbei are lower (Xu & Sun, 1996, Fen, 2000).

Under the influence of a multi-ethnic heritage, people of this region are generally famous for their hospitality and their loyalty to friends with whom they have personal connections. The practice of personal connections is characterized by building social ties with high levels of affection. Individuals are deeply embedded in social networks, so that sometimes it is hard to distinguish one's private life from public-related concerns. This is also shown by the way people address their counterparts in daily life. Age difference is a primary criterion that governs personal relationships, with others being addressed as relatives, such as 'sister' or 'uncle'. Personal networks are utilized on many occasions; and can also be adopted as strategy to exchange resources and exert power. In particular, the strength of ties between counterparts is crucial in exchanging interests. The stronger the personal ties, the more successfully they are used in various personal, economic and political interactions.

In this chapter, the Association of Saunders's Gull Protection (ASGP) is chosen as a case study. This environmental organization aims at protecting black-beaked gulls. Established in 1990, this association is a relatively old grassroots group. It has built up a certain amount of social leverage in Panjin locally, and around the country as well. There are few alternatives in choosing environmental NGOs in Panjin or other urban cities in that province, as very few environmental groups exist in Northeastern China.

The field work for this study was conducted twice, in May 2004 and August 2005. The triangle method was applied in this study to collect data, including face-to-face interviews consisting of open questions, documentary research and observation. Being a member of the research staff from the NGO Research Center at Tsinghua University, the author was accepted to conduct research with this group. She conducted interviews on leaders and members of ASGP, local officials, and ENGO staff in Beijing with whom the organization interacted. By a snowball method, individuals in personal networks were selected and introduced one-by-one in Panjin city. In this way, information was gathered on the functioning of the personal network. In addition, news reports, documents and existing literature, and interviews that have been published were also collected.

Protection of the Saunders's Gull

Background information on the species

The Saunders's Gull is a small seabird that has a black hood and nape and black beak, and is also called the 'black-beaked' gull. This species is in an endangered situation. According to calculations, in 2001 the total population of this seabird species was 7100–109,600 worldwide. The population is declining, according to projections, at an increased rate over the next 10 years.[8]

The Saunders's Gull is known to breed only in a few coastal areas in eastern China and South Korea. Two of the largest breeding colonies of this species are the Shuangtai Hekou Nature Reserve and the Yancheng Reserve in Jiangsu. Around 3200 of them were counted in 2001 in the Shuangtai Hekou Nature Reserve, which amounted to 90 per cent of the total population in Mainland

China.[9] The emerging population of the seabirds is dispersed along the coast of the Gulf of Bohai in Hebei and at the Yellow River delta in Shangdong. Birds breed at the Shuangtai Hekou National Nature Reserve from mid-March for eight months; and depart in mid-October. In winter, the Saunders's Gull's main passage and wintering is in the Russian Far East, Japan, South Korea, mainland China, Macao, Hong Kong, Taiwan, and Vietnam, which shows a linear distribution pattern.

The gull's habitat and breeding locations are almost exclusively along the coast. They built their nesting colonies in coastal salt marshes; and feed on exposed intertidal mudflats. Tidal flat and beach, therefore, provide crucial conditions for Saunders's Gulls, which are mostly found in wetlands.

Institutional setting of the Saunders's Gull protection

Despite their being a precious species, the existence of Saunders's Gulls around Panjin was discovered only two decades ago. Before the late 1980s, local authorities and inhabitants had no knowledge of black-beaked gulls. In 1986, experts from the national government and abroad visited and investigated the wetland reservation park for black-beaked gulls. For the first time, this precious species was known to have a breeding ground and habitat around Panjin. The number of Saunders's Gulls that bred in this area was estimated at around 1200, with the total world population at 2000. This fact did not have much influence on the local government's policy agenda when large-scale land reclamation was carried out in the late 1980s. Around 12 billion acres of wetland was utilized for agriculture, with one third being planted with rice. This caused land that gulls could live on to become very limited.

In spite of the poor contribution of local government to protect gulls, the central government established the national Shuangtai Hekou Wetland Reservation Park (STHWRP), also called the Shuangtai Estuary in the 1980s, aiming at protecting the precious ecological site. Bordering the Bo Sea in the East and farms of bulrush in the north, 30 kilometers south-west of the city, this reservation park covers 1280 square kilometers (Map 3.1). It covers the essential part of the world's largest reed marsh, a large area of the Suaeda community, and shallow sea. The site provides flood control and prevention; maintains groundwater recharge; and retains 10.4 million tons of nutrients and sediment each year. It provides also an important habitat for the resting and breeding of 400,000 waterbirds spanning 106 species.[10] The conditions in this nature reserve are quite favorable for Saunders's Gull: it is made up of several types of wetland—bulrushes, swamps, intertidal zone, shallow sea, rivers, dam, and paddy field. Bulrush amounts to 39.4 per cent of the total area; rivers cover 0.6 per cent of the area.[11] Such conditions are very rare around the world. From the perspective of wetland protection, although this site meets all eight criteria of the 'Ramsar Convention on Wetlands of International Importance', the Chinese government did not list the site until 2005.[12] One of the reasons for not joining this convention could be the exploitation of Liaohe

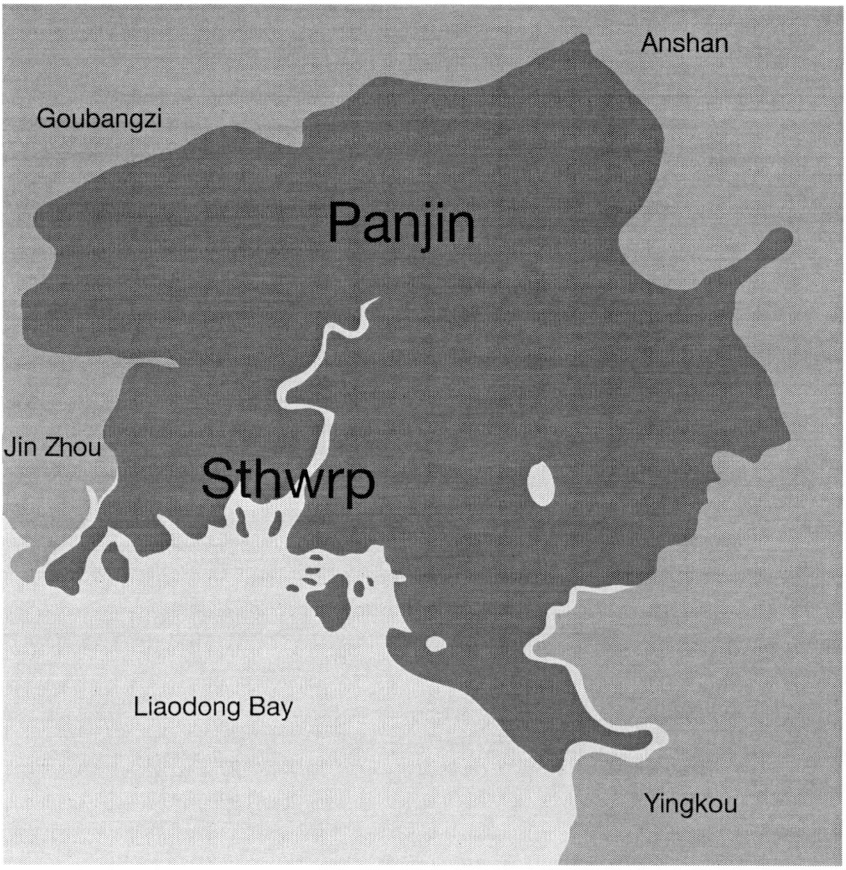

Map 3.1 Panjin and the national Shuangtai Hekou Wetland Reservation Park (STHWRP)

Oil Field, from which large profits were gained. Only in 2005, with the gradual depletion of this resource, was the STHWRP listed to become part of the Convention by the Chinese government.[13]

The government's actions were the main form of protection. Since their discovery, official protection of Saunders's Gulls was initiated under direction of the central government. The local Natural Resource Protection Bureau (NRPB) was established in 1987, a government unit directly in charge of wetland protection. Its responsibility includes the protection of not only Saunders's Gulls, but also the wetland, a rich source of various species.

Unfortunately, administrative regulations posed a major problem, as management and supervision of the natural park are separated (Figure 3.1). Originally, NRPB was established as a branch of the Ministry of Forestry in the central government. But this governmental unit is not financed independently by the central government, and is incorporated in the local political system.

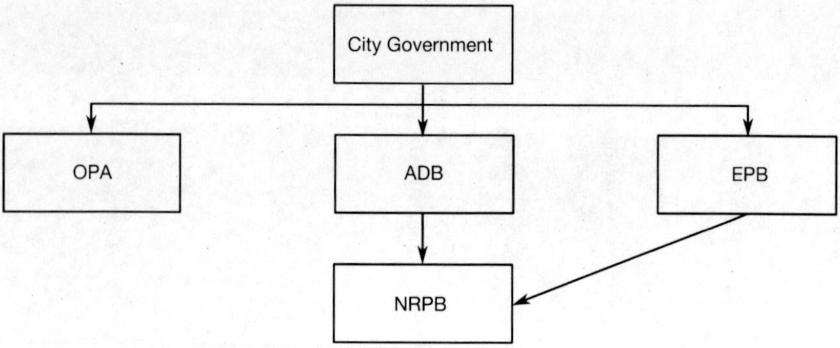

Figure 3.1 Administrative structure of NRPB under Panjin City Government's governance

In Panjin city, the NRPB has two supervising agencies. On the one hand, NRPB's work is under supervision of the local Environmental Protection Bureau (EPB), which is responsible for the protection of natural reservation park, but provides relatively general supervisions. On the other hand, NRPB is under regulation of the Agriculture and Development Bureau (ADB) of the Panjin Government. This bureau is responsible mainly for agriculture, forestry, livestock farming and reeds, and the protection of wild animals and natural resources. Compared with the EPB, the ADB provides more concrete directions and commands to the STHSRP. In the ADB's daily regulation, it provides financial income and administrative arrangements to the NRPB. Such arrangements leave relatively little autonomy in NRPB's work. Particularly when the protection of STHWRP is in conflict with local government's economic development activities, the interests of protection would usually be sacrificed. According to the 'Protection of Nature Reserve' regulations promulgated in 1994, companies are required to have consent from ADB before they exploit oil wells in nature reserve areas. However, individual companies did not comply strictly with this law.[14] Furthermore, because oil exploitation is of high importance for local economic growth, the ADB could hardly overrule the Office of Petroleum Affairs (OPA)—a government unit that belongs to the local government and plans oil exploitation. The illegal exploitation of petroleum might disturb the habitat of Saunders's Gulls.

In addition, the NRPB does not have complete authority over land use with regard to the bulrush farms that are in the charge of both Dawa and Panshan counties in their administrative regions.[15] Because the site of Shuangtai Estuary also provides seasonal employment for 20,000 people from these two counties for reed irrigation and harvest, the preservation of the site was also disturbed by the reclamation of land and the planting of rice. Between 1987 and 2002, the wetland in the nature reservation was reduced by 60 per cent.[16] Fishermen and peasants have the right to use the land in the nature reservation. Sometimes they collect seabird eggs for food, which probably leads to a reduction of the breeding of Saunders's Gull, especially as late-replacement clutches are vulnerable to storms. On the whole, governance in this area is concerned not only with the management of this park, but also with solving conflicts between various stakeholders who share land in it. Unfortunately, the administrative setting of the NRPB impedes good coordination, and disputes over land use of the STHWRP often happen.

The Association of Saunders's Gull Protection and personal connections

Founded in 1990, the ASGP is a grassroots voluntary group that aims at protecting black-beaked gulls. The main strategies of the ASGP comprise advocacy and education. In its advocacy activities, the ASGP focuses especially on the NRPB's policies and measurements related to black-beaked gulls and their habitat. Public education is another strategy that aims at promoting the public's awareness of Saunders's Gulls, which is mainly conducted among students in primary schools, middle schools and universities.

Viewed from the time it was established, the ASGP is probably one of the oldest ENGOs in China, although it always had a limited number of followers. Only in recent years, since 2000, has the association witnessed a growth in members, which have reached 500, around 120 of whom are core members who participate frequently. It has also recruited five schools as collective members,[17] including two primary schools, two middle schools and one university.

Internally, the ASGP is organized very loosely. Institutions are built with a low level of compliance. Lacking staff and personnel, its members also act as volunteers to work irregularly for the ASGP. It has an unstable, small budget of less than 50,000 RMB (approximately US$7500) per year. The source of income is composed of international foundation donations and membership fees.[18]

Organizational development

Since it was established, the ASGP has made little progress in its internal organizational development. The group has poor mechanisms to manage its members and to decide on internal affairs; the relationship between members is not regulated by rules and regulations.

The ASGP does not have a formal membership system. The procedure to become a member includes mainly registration and payment of membership fee. When mobilizing constituencies, Mr Liu, the leader of the ASGP, plays a crucial role. A majority of ASGP members were recruited through his personal connections with respective individuals. Particularly since the ASGP has a relatively low profile among the public, this was the main mechanism utilized to recruit members.

In the city of Panjin, besides relatives, friendship is one of the first steps in developing personal connections, which usually result in intimate and emotional relationships. Mr Liu is a person who has the charisma to build personal relationships. Being the leader and founder of the ASGP, his experiences in building the Association make an impressive story. He was the first journalist to report on the story of black-beaked gulls. From that time on, he has spent a great deal of his spare time in the wetland protection park to observe and investigate the gulls. After making great efforts to try to raise people's consciousness in protecting black-beaked gulls, he decided to establish the ASGP, in which he invested time, energy and quite a large amount of his own money. Under the socio-economic conditions of Panjin, where social voluntary groups are relatively unknown to citizens, Mr Liu's efforts in building a social association did not gain many positive responses. Most people, even his family, did not understand his efforts. For them, Mr Liu was wasting his time on irrelevant and uninteresting issues. Despite finding little support, Mr Liu persisted in his idea. His persistence leaves an impression of an optimistic, positive, and charismatic leader.

Mr Liu is also strategic and successful in developing personal relationships. He allows his counterparts to feel unique in their interactions. Potential members of the ASGP are treated as important guests, and are given gifts or invited to dinners. As a result, the connections that Mr Liu has built are characterized by a great sense of familiarity and intimacy. A kind of reciprocity is also formed that makes his counterparts feel compelled to do something for Mr Liu in return. Moreover, to be recruited as a member appears to be relatively easy.

Mr Liu's affection for nature strongly encourages and supports individuals who share this emotion. For instance, Mr Li, a retired official from a local state-owned company, loves wild birds and is fond of drawing them. He had an ambition to paint impressive drawings to educate people to care for birds. Mr Li got to know Mr Liu at his exhibition. From then on, he was warmly encouraged and supported mentally by Mr Liu in his drawing activities: 'His support helped me to go on.' Urged by Mr Liu, Mr Li joined the ASGP.[19]

The incentive that enables individual followers to remain members also depends on their personal relationship with Mr Liu. On the one hand, the bond of social ties is not easily broken, and may increase with time. The longer the connections last, the stronger the relationship becomes, and the more intensely the obligation to bring favors to the organization is felt. Thus people remain members for as long as their personal ties with Mr Liu endure. On the other hand, the benefits of being a member of the ASGP bring incentives for individual followers. Such advantages or gains increased since 2003, when the group began to gather sufficient resources to organize excursions. Also, members can represent the ASGP in receiving free training in Beijing, the capital, which is regarded as a nice opportunity to broaden one's perspectives. When Mr Liu has dinners with potential member of the ASGP, he usually would have his 'core' members accompany him, which helps to create a natural atmosphere. Such activities are usually free for members. The more an individual member can benefit, the more likely he or she will remain a member. But because the management of members is based on informal interactions, rather than a formal membership institution, the informality of the ties of individuals leaves little space for the establishment of formality. As a consequence, the efficiency of the internal membership organization is relatively low, and a very fluid structure of membership management exists.

Decision-making mechanisms

In the ASGP, the mechanism for internal decision-making is weakly developed. In the first 10 years of the Association, its poor working conditions further worsened this decision-making structure. The ASGP lacked facilities such as office space, communications equipment and common office facilities. Members would work in the leader's working place when he had his own office. These conditions lasted for around two to three years. Later, no office space was available and the core members met informally. The conditions forced the ASGP to operate in an informal fashion. Without a specific agenda being planned for meetings, members would gather irregularly in restaurants.[20] Usually, it was Mr Liu who initiated such gatherings. He acted as chairman on such occasions, which usually had a casual atmosphere.[21] Aside from meetings, there was no mechanism for the dissemination of information and the determination of internal and external strategies. Meetings did not take place as prescribed by the Association's constitution, according to which meetings for members to discuss organizational development and strategies must be held once every three years. In fact, informal discussions among members happened only on very special occasions. As the founder, Mr Liu acted as coordinator and main money-collector, and thus as natural chairperson of the organization.

The informal mechanism of decision-making did not mean there were no discussions or disputes among members, or opinions differing from those of Mr Liu. However, no strong opposition was or is made against Mr Liu, for the fear of losing personal ties.[22] The value of personal connections was believed to be so important that no actions or initiatives were taken that would endanger these connections between Mr Liu and individual members.

Internal relations

In the ASGP, personal connections are especially strong between Mr Liu and the members. In comparison, the personal relationships between members are not as close as they are with Mr Liu. He is crucial in sustaining the unique personal connection that bonds members to the ASGP. In such a context, it is questionable whether the ASGP will remain as it is, if Mr Liu is no longer chairing. His absence would probably jeopardize the ASGP's network and result in the loss of many members.

Members of the ASGP hardly ever feel a sense of belonging or solidarity related to the collective.[23] Even their goodwill in protecting black-beaked gulls does not form a strong enough bond of solidarity between them and the organization. For example, the ASGP does not have a website. Neither does it have regular publications or widely distributed communications. During the 12 years of its existence, only a few pamphlets have been made as journals, in 2001 and 2002. After that, no publications have been produced. The ASGP suffers from a lack of visual devices that could solidify its existence and create an identity for its members. This void strongly affects young members who highly value such an identity. In 2002, some students from Beijing voluntarily contacted the ASGP after hearing about it. They were enthusiastic in participating in the Association as members and working for the protection of black-beaked gulls. However, in the entire year following their registration, they heard nothing from the ASGP. This disappointed these young students greatly, and they later left the Association.

Composition of membership

The members are a fluid group with no specific collective identity and do not belong to any particular social background. Individuals who possess resources that can be useful to the ASGP are persuaded and mobilized by Mr Liu. Various materials are considered useful resources: money, connections, and channels that can be used to influence the government or to promote the ASGP in having a stronger social influence. The organizational relations of the ASGP with other institutions are also built through Mr Liu's personal ties with key figures. According to him, this is a strategy to get things done. The association would be like a family, with its participants being like brothers and sisters to each other.[24] When key individuals from other organizations join the ASGP, it becomes easier, through personal interactions, to communicate with and persuade these organizations. Furthermore, chances exist for the ASGP to influence politics.

The ASGP's members can be divided into three parts according to their occupation: media, government, and culture- and education-related occupations (Table 3.1). Among the current 120 core members, 41.5 per cent have a media-related job, which forms the largest group of members. This is a natural result stemming from the fact that for the past 20 years the leader himself has been working for a newspaper, which is very much reflected in his personal

Table 3.1 Composition of members

Occupation	*Percentage of the members*
Civil servant	25
Journalist, photographers and other media	41.5
Teacher	12
Employee in industry	11
International expert	4.1
Others	6.4

Source: data provided by ASGP, September 2005.

network. Mr Liu and his colleagues from the largest local newspaper, *Panjin Daily*, form the most important segment of the core members. The *Panjin Daily*, a newspaper that reports particularly on the city of Panjin, covers many issues regarding politics, society, and economy. Its daily circulation gives it an advantage by creating more opportunities to publicize the ideas of the ASGP. This newspaper, although it belongs to the local Chinese Communist Party, has a surprisingly open attitude. It has a considerable degree of independence, and feels responsible for the promotion of social improvement, for instance by raising environmental awareness.[25] This is shown especially in the attitude of the newspaper's management toward Mr Liu's great amount of work on black-beaked gulls and nature-related issues. His work is valued highly, and following his work on the wetlands he was promoted to 'Journalist Specialized in Ecological Reports'. This support provides Mr Liu with greater freedom and better control of news reports on ecological issues in this newspaper. Members of the ASGP also come from other local or provincial media agencies, including newspapers, radio, and TV stations. Such liaisons with other media agencies contribute to the capacity of the ASGP to utilize mass media to publish and disseminate its ideas.

Aside from media personnel, 25 per cent of the members possess a middle or higher position in the government, which constitutes the second group of members in the ASGP. It has always been Mr Liu's deliberate purpose to get officials involved in the ASGP, as these individual officials will be easier for the Association to approach. As member of the ASGP, their opinions on environmental protection issues, and thus government decisions, can be influenced (Table 3.2).[26] By inviting higher-ranking officials to act as honorary leader of the ASGP, connections were built between the ASGP and respective government units.[27] Through these relationships, the ASGP has mobilized useful resources such as political influence, subsidies, office space, transportation means, and information.

The third largest portion of members is formed by teachers from different educational institutions. This group is important in the education of pupils and in helping young people to become involved in black-beaked gull protection. Due to these individual members, two primary schools and two middle schools have become collective members of the ASGP. Young children began to know about black-beaked gulls, and their awareness of nature protection was raised.

Table 3.2 Ranks of civil servants

Rank	*Percentage of total number*
Lower	57.5
Middle	23.3
Higher	19.2

Since 2000, personal ties have also been built with heads of enterprises. Through these private sector members, Mr Liu has facilitated collaboration to develop environmental education, including financial support of money and facilities.[28] However, cooperation with these private counterparts has just begun, and remains relatively limited.

Organizational relationship with government

As analyzed in the first part of this chapter, due to the political situation in Panjin, the space for volunteers and activists in protecting black-beaked gulls is very limited. There is a lack of recognition from both society and the political system. For quite some time, the ASGP had no legitimacy in conducting black-beaked gull protection activities, which was believed to be the government's responsibility. Members of the ASGP are very aware that without the government's involvement, it could hardly initiate and successfully implement its activities. So building organizational links with political authorities became the main task of the Association from the early days onwards. The Association's main tactic in interacting with political institutions was to cultivate personal connections with officials of different rank and from different government units. The relations between the ASGP and the government were mostly at an individual level—between the leader of the ASGP and specific political elites. Through personal contact, authority and influence were gained.

Interactions in early stages of the ASGP's development

Construction of personal ties is based on delicate estimations and calculations. In the first members' meeting, held in 1990, the leader of the ASGP set out a strategy to persuade a Minister—Mr Pz Liu from the Ministry of Agriculture—to become the honorary director of the ASGP and to participate in the gathering. The old man was an influential figure in Liaoning Province. Mr Liu carefully wrote a letter to Mr Pz Liu and prepared a poster of a Saunders's Gull for him, outlining the vision that motivated him to establish such a voluntary group. His passion moved the old man, who agreed to attend the first members' meeting in 1990. This tactic of cultivating relationships with influential and high-ranking political elites was also used to assemble officials from the local government. Being very aware that the principles of personal connections work in a similar way in the government, Mr Liu estimated that officials of lower rank would also participate in the members' meeting, to show *mianzi* (face) to their boss.[29] When higher-ranking civil servants consent on something, the lower-level officials usually agree with their supervisor. The consent of Mr Pz Liu constituted one important step in developing wider relationships with government officials. His presence to a large extent affected and motivated the top officials in Panjin to participate in the following years. In this way, through Mr Pz Liu's social prestige, the ASGP successfully arranged the support of more than 20 officials, including the heads of the Bureau of Agriculture, the ADB, and the NRPB, officials from the county of Dawa and Panshan (which are also related to the protection of the STHWRP), and others. By getting these officials involved, the ASGP gained authority and influence, and became well recognized by most governmental agencies related to wetland protection.

In the early years, Mr Liu interacted primarily with the NRPB. Personal relationships established between Mr Liu and officials of this agency did not lead to organizational links. Instead, they were utilized by officials of the NRPB for their personal interest, thus links between the two political organizations largely stayed at the individual level. In the relationship-building between Mr Liu and the officials of the NRPB, Mr Liu had to conduct very close personal contacts with specific leaders. In order to create working relationships, a certain level of intimacy was established between them, for instance they called each other 'brother'. During Mr Liu's efforts to link the two institutions, he was requested for various favors by his counterparts. Mr Yang, Head of the NRPB, asked Mr Liu to work for his personal interests. Knowing that Mr Liu had wide contacts with journalists, he asked him to contact media personnel to help build a good public image and raise political credibility for himself and his organization.[30] Aside from building links with the media, Mr Yang also consulted Mr Liu about policy arrangements on wetland and gull protection. Mr Liu was known for his knowledge on wetlands, which resulted from years of investigation and knowledge accumulation, and he was invited for policy discussions on the opportunity to develop tourism in the wetlands. However, he was invited not as leader of the ASGP, but rather as a journalist and as an expert on black-beaked gulls.

So the particular individuals, rather than the profiles of their organizations, determined the development of these relationships. Likewise, the maintenance of the relationship between the organizations was affected by the stability of personal ties. Mr Zhang, the current director of the NRPB, was promoted to his position in 2002. Because of the position he held in this agency, he was invited to become honorary director of the ASGP. In 2002, a reporter who worked at a local TV station made a documentary about black-beaked gull protection in Panjin city, entitled 'The Saunders's Gulls Protection between China and Japan'. Mr Liu was portrayed as the main person in this documentary, representing the public's efforts in protecting these precious species. Mr Zhang was given only minor attention in his position as an official.[31] The documentary won the first prize in a provincial documentary competition, providing Mr Liu with a strong reputation. Mr Zhang, however, noticing that Mr Liu gained all the merit, was not satisfied with the result.[32] His relationship with Mr Liu then changed dramatically. It deteriorated further when the ASGP organized a conference on the gulls' protection later that year. Participants included not only domestic experts from other wetland protection reserves, but also those from Korea and Japan. Mr Zhang, as representative of the NRPB, also attended the conference. But he kept the conference papers within the NRPB, preventing Mr Liu and the ASGP from publishing them, and thus from achieving even more credit.

In order to safeguard future favors from the NRPB, Mr Liu continued communication with the directors of the NRPB. His efforts to support officials in the NRPB led to restricted forms of collaboration between the two organizations that shared the same goal. However, in the past ten years very limited collaboration has been realized. Personal networks with the directors of the NRPB were characterized by unstable and temporary relationships, which depended strongly on the personalities of individual directors.

Interactions from 2002 onwards

With relatively weak organizational relationships between the ASGP and the NRPB, the ASGP remained weak in its influence on policy and other political matters. Transitions in the organization's development occurred only in recent years. This was caused through the construction of a link with the Secretary of the Party Committee, Mr Zeng,[33] who highly valued environmental improvements in local governance. He advocated policy plans to construct Panjin as a more ecological city. He had ideas to control water pollution, to adjust the economic structure, and to promote ecological agriculture. Under his influence, a new political agenda that focused on sustainable development was developed in 2002. But this policy shift was new to the existing political leaders. Although it indicated a direction for future development, the plans and ideas were not always very concrete and clear. It resulted mainly in an agenda-setting process, while implementation was realized in limited areas, such as in food and the plastics industry.

However, in the eyes of Mr Liu, the political ideas raised by Mr Zeng appeared to be an opportunity for the ASGP to raise its profile and influence. With no other institutional channel available to seek resources and support from government, Mr Liu wrote a letter to Mr Zeng to introduce the work of the ASGP and its poor working conditions. A reply was soon received, starting a process of trust-building between Mr Liu and Mr Zeng—the latter used his power to provide an office in the local EPB. In return, Mr Zeng accepted the offer from the ASGP to be its honorary director, which brought Mr Zeng political credibility. This title is highly valued by the political elite, as it shows that a civil society organization appreciated his work. Zeng's publicity gained attention from the public as well as from provincial officials. He was promoted to the provincial government as a Vice-President of the Party Committee in 2004. From that time on, he did not show any interest in, or give any direct support to, the ASGP, illustrating the strategic use of personal connections.

Regardless of how short the involvement of Mr Zeng was, his participation encouraged members of the ASGP to become more active and more visible by the government. A public ceremony was organized when Zeng was awarded the certificate of honorary director. Despite the content of the ceremony being simple, it gained large publicity at that time. The participation of Mr Zeng brought along other officials, in order to give him face. Six main government officials attended the ceremony, such as the Vice-Secretary and the Vice-Mayor of The City Government; two top officials who worked in the local

People's Conference; and a Vice-Director of local Chinese People's Political Consultative Conference. Around 25 ASGP members, young and old, came and brought their songs and poems. Along with these participants, around 20 journalists, photographers, and TV personnel were present. The ceremony was broadcast live on TV. The most visible moment in the ceremony was the donation of an expert certificate to Mr Zeng. The fact that an ordinary person endowed a certificate to a top official strongly impressed the Panjin people. For many, this was their first encounter with a social organization. The high official's involvement reflected the ASGP's legitimacy, endowed by the government, and enabled it to gain a stronger reputation and acceptance among the public.

Since 2004, the ASGP's involvement in policies has been strengthened due to a link with another political official, Mr Js Liu, the Vice-Mayor of the city responsible for environmental governance. Following their wives' personal relationship, the political leader was connected with Mr Liu, and started to support the ASGP. Subsequently, Mr Liu was invited to government meetings on issues related to wetlands. This proved to be a major step in participating in local environmental governance.

Thus the establishment of personal ties with political elites has proved a useful strategy for gaining resources and support. The informal interaction between the leader of the ASGP and political elites has had a strong impact on the ASGP's social leverage. The political elites with whom Mr Liu interacted brought more influence to the ASGP. The ASGP's influence has been broadened and deepened, especially after 2003 (Table 3.3). It has challenged NRPB policies in recent years more than in the 1990s.

Consensus building through mass media

Consensus-building is another important strategy used to enhance the ASGP's influence. Directed by Mr Liu, who knows the media's limits and boundaries, the utilization of newspapers went very smoothly. Regarding the content of media campaigning, the ASGP targeted both public and government authorities. It established a discourse protecting Saunders's gulls; and built legitimacy for

Table 3.3 Influence of personal connections with officials

Items	*1990–99*	*2000–02*	*After 2003*
Counterparts	NRPB	NRPB	City government and EPB
Length	Not stable	Not stable	Relatively stable
Nature of connections	Relatively good	Not very good	Good
Result	Collaboration on both organizational and individual levels	No collaboration on organizational or individual level	Collaboration on both organizational and individual levels

the ASGP. Furthermore, it mobilized the media to improve policy measures to protect black-beaked gulls. Through the years, this strategy of building consensus saw a gradual change from mild towards more radical, along with the social leverage that the ASGP gained.

Consensus on protecting Saunders's gulls

Constructing 'public consensus' is important in a society where the public strongly believes in governmental authorities. Specifically, in Panjin, public framing proves to be useful for educating the public, who in general had very little knowledge about black-beaked gulls and their endangerment in the early 1990s. The media are a natural resource for the ASGP, because a majority of its members are related to the media. Mr Liu capably built advocacy through publicizing articles in the *Panjin Daily*, which was the main public framing activity of the ASGP for a long time. Through collaborating with the media, the ASGP framed the issue of the black-beaked gull and raised awareness of its protection.

Publishing personal stories proved a successful way of enhancing the public's attention on this special species. A typical story was on the saving of a small, dying black-beaked gull by the members of the ASGP.[34] It describes, at length, the care and love of these members in nursing a little dying black-beaked gull, which was finally saved after two weeks. This story was also reported by many provincial news media. Later that year, the members were invited to Beijing to be interviewed by a national TV program on this matter.[35]

Newspapers and the media were also a key element in building consensus on the importance of black-beaked gulls, and thus of the ASGP. In 1995, a public discussion was organized in the *Panjin Daily* for selecting which bird represents the city. This produced a large response among the public, and received attention from the city government. A specific committee for 'selecting the bird of the city' was established to deal with this issue. During the four months of public discussion, Mr Liu publicized around 20 articles introducing the details of black-beaked gulls.[36] Words, charts and pictures were used to introduce the various aspects of gulls, as well as their scarcity. Therefore, while the introduction to other kinds of bird remained rather limited, the public was familiarized with black-beaked gulls in full detail. After a four-month discussion, both black-beaked gulls and red-crowned crane were chosen to be the symbolic birds of Panjin city.[37]

The Association's construction of legitimacy

In the process of dissemination, not only articles about black-beaked gulls were publicized, but the ASGP was presented as well. In articles about black-beaked gull protection, the ASGP was often mentioned to make its name and activities familiar to the public. Also, beyond this context, Mr Liu carefully arranged bird stories and introduced the ASGP in other suitable articles that have news value.

In order to enhance publicity and the legitimacy of the ASGP, the group initiated some activities to catch the public's attention. In July 2002, a ceremony was organized to enhance the public's consciousness in protecting black-beaked gulls. The theme of this ceremony was expressing sentiments and protecting the bird's homeland, which is endangered by economic exploitation due to petroleum withdrawal. Not only students and young people attended and performed by singing and reading poems, but also government officials from the EPB, the NRPB and the OPA, whose jobs were related to protecting the wetlands. In this ceremony, sentiments of love and care for black-beaked gulls were expressed. The program included setting a recovered gull free and praising the members who took special care of it. Poems written by Mr Liu, leader of the ASGP, were read, including such titles as 'Go Back Home, Little Gull', and 'Mother Please Do not Worry'. This activity was reported in the *Panjin Daily* several days later.[38] Other articles, including news briefs that could increase public's recognition of the ASGP, were often written, such as an investigation conducted on the ASGP by a leading Chinese university studying social organizations. This particular piece of news appeared as a 300-word news brief on the first page of the *Panjin Daily.*[39] In this article, the fact that this group had caught the attention of a leading national research institution was emphasized, acknowledging its organizational significance and legitimacy.

Through these media activities, the ASGP's leader, Mr Liu, gained recognition as an influential person and an expert in black-beaked gull protection. Stories on him protecting black-beaked gulls appeared in local news media, and also in provincial and national media.[40] This brought Mr Liu fame and reputation. Due to his publications and the activities he organized to protect black-beaked gulls, he was honored with national environmental prizes. The prizes and fame created Mr Liu's image in public: he was celebrated as an 'expert', an 'authority', the 'father of black-beaked gulls', and the 'god of black-beaked gulls'. The position of Mr Liu is more than a personal achievement; it also represents the social leverage and progress made by the group as a whole. ASGP's reputation was therefore facilitated and enhanced.

Campaign promoting strict regulation measures

An important media campaign was organized by ASGP in 2003. It was triggered by a reduction in the amount of black-beaked gulls from 5020 to 2930, as reported by the yearly counting in 2002.[41] Facing this critical loss of gulls, Mr Liu designed a media campaign on this issue. At first, he let a journalist who is member of the ASGP publish an article reporting the situation. The published article had a very alarming title—'The Death of 2,000 Saunders's Gulls'.[42] According to the ASGP, the lack of wetland management and expansion of the agricultural area were the main cause of this dramatic loss. The NRPB, as main wetland manager, did not have full control in keeping the wetland area intact, and villagers could enter it without difficulty. As described in this article, villagers stole gulls' eggs and disturbed the habitat of black-beaked gulls. Several days later, Mr Liu published another article, entitled 'The Crying Xiaonan River'.[43] The Xiaonan River was the area where black-beaked gulls used to live. He blamed the loss of black-beaked gulls—again—on weak control by the NRPB, and the concentration of too large a number of gulls in this area of only 28 square miles. According to the report, this phenomenon started in 2000, when the expansion of agricultural areas forced the gulls to live in a small area near the Xiaonan River. Later in 2003, flooding of the river led directly to the deaths of many young gulls. Adult gulls fled their home, leaving young birds dead in their nests. In his article, Mr Liu emphasized that this river was the last area where black-beaked gulls could live peacefully, and he advocated stringent protection of this area. In a grave and sad tone, he criticized the lousy attitude of the NRPB's officials. But, according to NRPB officials, this loss was a natural consequence, which was not very serious: 'new ideal places for black-beaked gulls will be available soon'. Mr Liu did not agree, and insisted that the future for Saunders's Gulls was dangerous, and that measures should be taken.

Both ASGP articles were lengthy, around 3000 words. After these publications in the local newspaper, stronger pressure was developed by a further publication in a national newspaper.[44] This article alerted the leader of Liaoning Province, who was convinced by the ASGP's ideas, but also by the potential political repercussions for the image of his province. He demanded measures to be taken in order to save the black-beaked gulls. The NRPB was reprimanded for its careless attitude, and ordered to implement the political leader's request. Around 56 square kilometers were reserved specifically for Saunders's Gulls as permanent breeding place.[45] In this way, ideas from activists were realized.

Thus building consensus has proved an important and successful strategy of the ASGP. However, its influence is often limited, and it is only effective in bringing specific issues to the attention of the wider public in the short term.

Epilogue

The efforts that the ASGP made in Panjin city brought forth a quiet environmental movement. It raised public environmental awareness on protecting black-beaked gulls, and had impact on policy arrangements. The most concrete improvement is that the number of adult black-beaked gulls in the Panjin wetland has reached 5699 in 2005, according to the NRPB.[46] This is around three times that of two decades ago, when the ASGP was founded. Furthermore, the ASGP is now to some extent involved in environmental discussions with the local EPB, which shows its growing role as an actor in local ecological governance.

This case study indicates that ASGP shows steady organizational development. Yet, 15 years after its establishment, its level of institutionalization remains low, which largely follows the utilization of personal networks. Internally, the ASGP relies on interpersonal relationships for communication, coordination and decision-making. The vulnerability of individual ties jeopardizes the (institutional) development of the group. Little formality exists between members, and hardly any collective identity could be constructed among them. Values and ideas that are shared among participants in the Association and the leader are based on their personal interactions, instead of the group's common identity. Regarding the prevailing adoption of individual ties in the ASGP's development, we can conclude that using personal ties is one of the dominant strategies of environmental organizations in Panjin. Viewing its development over the past 15 years, it appears that this strategy of personal networks proved to be of more importance for the ASGP than the internal institutionalization and formalization, and the external contribution of organizational networks.

4 Environmental activism in Beijing

Beijing's socio-economic context

Environmental governance in Beijing

Beijing, the capital of China, has a population of 15.3 million.[1] This city has witnessed fast economic growth in recent years. An average 10.3 per cent increase in economy has been achieved between 1979 and 2004. In 2005, the average GDP per capita reached US$5457, which makes Beijing one of the richest cities in China. This average income per capita has reached the level of medium developed countries.[2] Yet Beijing does not have a positive environmental record. It has been seriously polluted and the air quality is low, caused by industrial pollution, coal combustion, and transport emissions.[3] Its surface water is also polluted, and the city faces serious depletion of underground water.

Only towards the end of the 1990s did Beijing's municipal government begin to take serious measures to improve its environment. In the past few years, Beijing's air and water conditions have improved greatly. In 2004 the airborne emissions of SO_2 have decreased by 9.8 per cent in comparison with 2003. The air quality also improved. In 1998, on 100 of the 365 days, air quality levels were above the national air quality standard; by 2004 this number increased to 229 days.[4] Concerning the water environment, the city has improved wastewater management and the protection of drinking water. Among the 76 rivers, only 23 have reached the water quality standards. In addition to air and water issues, the authorities also focus on solid waste management and noise pollution. Furthermore, projects protecting regional ecology and biodiversity are being constructed in order to realize long-term pollution control since 2004.[5]

In general, the improvements in protecting the environment are closely related to municipal government's determination to clean the environment. For years, the capital has been challenged by serious ecological degradation. When Beijing started preparations for hosting the Olympic Games in 2008, cleaning the environment became a priority on the municipal government's political agenda, and it thus invested a considerable amount of money in doing so. During the years 1999–2001, over 100 billion yuan were spent on this initiative. Since 2002, the amount invested accounts for 3.3 per cent of the annual income of the municipal government, which reached 141 billion yuan in 2004.[6] Furthermore, the city government takes strict measures to control industrial pollution. For example, one of Beijing's largest industrial companies—Shougang Group—has been relocated outside Beijing in order to reduce air pollution in the city. This move was a huge trade-off between protecting the city's environment and realizing its economic growth. Shougang Group constitutes a large proportion of Beijing's income, reaching approximately 80 billion yuan annually in the past few years. This company also provides tremendous work opportunities for the Beijing population—one in six of the industrial workers in Beijing work in Shougang Group. It is estimated that this initiative cost 500 billion yuan, and relocated 80,000 employees. From an ecological viewpoint, this initiative will significantly decrease the amount of particulate matter in the air per year.[7]

Besides administrative measures, the authorities in Beijing also demonstrated an open attitude by involving environmental NGOs (ENGOs) in environmental governance as a supervising force. The involvement of Beijing environmentalists is the result of the cooperative relationship between central government officials and ENGOs. The leaders of the Ministry of Environmental Protection (MEP), such as its former Head, Mr Xie Zhenhua[8]and the Vice-President, Pan Yue, have expressed their support to ENGOs in many formal speeches and conferences. Informally, these two officials, especially the latter, Mr Pan, are familiar with a number of NGO leaders, often meeting them on informal occasions.[9] In recent years, continuous efforts have been made by MEP to improve the status of ENGOs in environmental impact assessment (EIA). This is also related to the agency's efforts to gain influence over economic agencies in central government, such as the National Development and Reform Commission (NDRC). In the past, MEP's authority was largely ignored in EIA. The NDRC was responsible for the construction of projects that caused serious environmental degradation (Zheng, 2001). In January 2005, MEP halted 30 construction projects, and required another round of EIAs to be made. This action set off a campaign to strengthen the power of EIA,[10] and was supported by 56 ENGOs, which published a collective letter claiming partnership with MEP. Since then, MEP accelerated its promotion of ENGOs in institutional settings, as a contribution to environmental pollution control and representing the public's environmental rights. A regulation labeled 'Provisional Measures for Public Participation in Environmental Impact Assessment' was promulgated in February 2006. It is the first legal document that enables the participation of the public in environmental policy-making processes.[11]

Beijing's special political environment also provides room for ENGOs. Two political systems, the central government and the municipal government, are both located in the city. These two administrative systems have responsibilities in governing public affairs in Beijing that are not strictly defined. These deficiencies provide chances for ENGOs to initiate their activities in Beijing.[12]

Local culture and personal network

Built as the capital of the Jin dynasty in 1153 BC, Beijing is a city where both traditional and modern cultures meet. A mixture of both influences and ideas can be found in contemporary Beijing. Being the political center of China for 700 years has significantly shaped the city's local cultural characteristics and the role *guanxi* play in modern life.

Historically, Beijing has been the capital of three dynasties: namely Yuan, Ming and Qing. It has also been a key place where diverse cultural genres were seen, aside from being a political center. Its central position makes it influential in affecting other cities in the country (Yan, 2005). This city has thus brought its local inhabitants pride and confidence since ancient times. As one of the first cities that had contacts with international community since the seventeenth century, Beijing is also relatively open to receiving external cultures (Yan, 2005). As a result, compared with people from other places, its citizens are more polite, confident, and calm. In modern times, Beijing has developed to become a center with various subcultures, as diverse populations from other places across the country pour into the city. Yet the traits of local inhabitants are still prevalent, and distinguish them from people of other places in the country.

In daily life, Beijing people are fostered to adore politics and politicians, and tend to dislike business and merchants. Local inhabitants love to become officials, and in comparison, a sense of dislike of businessmen and merchants is formed, who are thought only to be after profits. This subculture values social ties higher than business contracts (Yang, M., 1994), with a tendency to ignore regulations and rules. It also developed into a strong sense of loyalty to friends who are linked through strong social ties. And utilizing personal connections becomes a way to withdraw resources and exchange interests. To a large extent, informal networks and personal ties are incorporated in formal institutional operations. All in all, using personal ties plays an important role in the local culture of Beijing.

Beijing's ENGO community

Historical review

Beijing's ENGOs were among the earliest to emerge across the country. A number of independent environmental groups exist. The areas in which these groups are active include environmental education, research and technological studies, and provision of legal service to environmental victims. When reviewing the growth of ENGOs, it can be seen that the community's growth is closely related to the personal relationships between group members. The changes in individuals' interactions further facilitated the splitting of old organizations and the formation of new ones. Using organizational growth as a meter, two stages can be divided.

The first emergence of environmental groups in Beijing is largely due to the growing number of environmentalists in the past decade. As analyzed in Chapter 1, Chinese environmentalists were greatly motivated by China's globalization process. The Rio Declaration, following the 1992 UN Conference on Environment, particularly enlightened Beijing's environmental activists. The first national grassroots ENGO—Friends of Nature (1993)—was established in such a context. Later, Ms Wang Yongchen, once a member of Friends of Nature, established Green Environmental Volunteers (GEV) in 1997. Along with Global Village Beijing (GVB), established in 1996 by Liao Xiaoyi, these three groups appeared as forerunners in organizing NGOs and mobilizing environmental protection among the public. They constitute the main groups in the 1990s.

These three groups have three features in common. Looking at leadership, the three groups all had charismatic leaders, who were capable of mobilizing resources and constituency. As far as the goal of the three groups is concerned, they all were committed to raising environmental consciousness through work at grassroots level. Finally, we see that the level of institutionalization of all these three environmental groups was low—they lacked internal institutions, a membership system and democratic decision-making mechanisms.[13] By the end of the 1990s, these three groups had produced certain social leverage in raising environmental awareness in Beijing. Enlightened by their educational activities, MEP also began to strengthen its measures in the field of environmental education, which resulted in the establishment of the Center for Environmental Education and Communications in 1998.[14]

In the second period, after 2000, another tide of ENGOs can be identified, with a majority of the ENGOs organized by former staff members of the previous groups. The founders of these groups were usually capable project officers. The fact that they possessed certain resources was one important reason for establishing their own groups. At the same time, most of these individuals also had strong disagreements with their leaders relating to organizational development, management, or project operations.[15] Personal tensions and dissatisfaction with current leadership styles became a main source of disagreements, which further led to splitting of organizations. The new groups established by capable staff usually focused on similar activities to the old ones.

At this stage, the establishment of the following new groups was of particular significance to the ENGO community in Beijing: the Beijing Brooks Education Center (2003) promotes citizens' education, especially focusing on peasants' education; WildChina (2001) is committed to protecting biodiversity, especially wild animals, through photography; the Beijing Earthview Environment Education and Research Center works on environmental education and related research; Shinestone Community Participation Action (2002) promotes citizens' participation in urban communities; the Global Environmental Institute (2003) focuses on cooperation between NGOs and enterprises, promoting the utilization of technology in protecting the environment; and the Green Cross (2003) conducts environmental education in villages. Some other new autonomous NGOs also emerged, such as Green Island, a group committed to raising environmental awareness among journalists and young people (2002); Desert (2002) is concerned with desertification problems; and Green Web (2002) is an internet-based volunteer group. These groups have a more professional profile than their forerunners. For instance, their target audience is more focused, as opposed to the mobilization of a large group, such as journalists (Green Island), and community citizens (Shinestone Community Participation Action).

Students' environmental associations compose another important part of Beijing's new ENGO community. Many of the young environmental activists were enlightened by a well known ENGO among university students, the University Green Camp. This organization, also called 'West Point' by environmentalists in China, is funded by an influential environmental activist, Tang Xiyang. University Green Camp has trained groups of young people and prepared them as potential leaders in ENGOs (Tang, 2004). University Green Camp is popular among university students, and provides opportunities for them to broaden their horizons and gain meaningful experiences. Under its influence, recent years have witnessed the establishment of dozens of student environmental protection associations in most of the universities of Beijing.[16]

Organizational links

Beijing's ENGOs are generally willing to cooperate with each other. They are aware that they can benefit from cooperation, especially when they are relatively small and have limited social influence.[17] Various degrees of coordination exist, such as sharing information, organizing workshops, launching programs collectively to achieve stronger influence, and so on. This tendency to collaborate is relatively strong, and distinguishes the ENGO community from NGOs that work in other areas.[18]

In fact, individual environmentalists are familiar with each other within the activist community, especially the leaders and key figures, who have many opportunities to interact, such as in common conferences, training courses and meetings.[19] As a consequence, cooperation becomes relatively easy. The ENGOs that were founded in the first period play an important role in the organization of cooperation through conferences, workshops and discussions. They are the most capable in such activities, and their influence facilitates their success in organizing collaboration. GVB organized a team representing Chinese ENGOs to participate in a 'Sustainable Leaders' Conference' in Rio in 2002. Friends of Nature, in the same year, organized a second assembly of the Global Environment Facility in Beijing, and created possibilities for Chinese ENGOs to attend. In this way, the exchange of information and experiences is often visible in the ENGO community.

In order to achieve more influence on the government, ENGOs have also formed alliances in recent years to organize campaigns. In 2001, a project constructing cement on the riverbank of the KunYu River raised protests from the ENGO community, which claimed that the natural ecology of the riverbank might be endangered. GEV, Friends of Nature and the Institute of Environment and Development collaborated with environmental activists to host a public hearing. Experts on ecology, inhabitants along the KunYu River and the Vice-Mayor of Beijing were invited to attend. This public hearing was stopped mid-way by the authority of a top official and prevented from being reported, and the project was implemented as planned. Nevertheless, the environmentalists demonstrated their potential and force in participating in environmental policy-making, and were greatly encouraged by this debut. Later, GEV was also involved in leading campaigns to protest against the Beijing Zoo's relocation plan, and against dam construction in Yang, Liuhu in Sichuan Province. Building collective campaigns has become one means through which Beijing's ENGO community exerts a certain amount of power and influence on environmental policies.

However, difficulties exist in constructing cooperation when the number of actors is large. Personal issues and dissatisfaction with current leadership styles are important elements hindering cooperation (Qian, 2006). The fact that ENGOs are familiar with each other does not ensure cooperation will succeed each time. The possibilities for forming an alliance largely depend on the recognition and reputation of group leaders.[20] The nature of individual environmental activists strongly affects the possibility of collaboration. If a leader of an ENGO is not appreciated, it may be possible that this group will not be incorporated in collective actions.

The influence of international NGOs

INGOs contributed substantially to the success of Beijing's ENGO community. Among the international support that local ENGOs receive, the USA, European Union and Hong Kong Special Administrative Region are the most prominent. INGOs foster the development of individual ENGOs and support them to operate in a professional way.

Support from the international community can be seen in several forms. First, INGOs directly provide funds to Chinese ENGOs. These INGOs are usually groups with relatively large annual budgets that have established their own office and staff in China, such as Friends of Nature (Hong Kong), the Ford Foundation, Partners for Community Development, and Oxfam (Hong Kong). Their general goal is to foster the development of Chinese civil society. ENGOs established in the first stage relied especially on these funds to establish offices and initiate their projects. In general, the funds provided were relatively substantial, often meant to cover administrative operations, and the staff of these environmental groups could live from the fund.[21] In this way, ENGOs in Beijing could offer their staff competitive salaries and became attractive to graduates and job-seekers as a career. Supported by foreign funds, a career in Beijing's ENGOs gradually became a promising opportunity for graduates.

INGOs also promote the formalization and professionalization of ENGOs. During the process of localization, INGOs recruit capable individual activists to act as their agents, and work in a formalized environment for local Beijing ENGOs. Since these individual activists have personal networks themselves, it is through such networking that professionalized management is taught and introduced to local ENGOs.[22]

Second, relatively small-sized INGOs recently sought collaboration with Chinese ENGOs, such as the Wild Flower Institute and Ecologia.[23] Due to their limited budget, they can hardly establish an office in Beijing, thus they choose to build partnership with local environmental groups and develop projects in collaboration. In this process, the Chinese partners have to operate in many aspects with their counterparts, and are forced to manage their projects professionally.

What is more, INGOs also encourage Chinese ENGOs to get in touch with the world outside China. Beijing's environmental activists are provided with a large number of opportunities to communicate with INGOs in developed countries and regions, for instance in Hong Kong and Taiwan, the USA and Western European countries. These precious learning experiences have enlightened many environmentalists. It was after such a trip to visit women's NGOs in the USA that Ms Wang and Ms Jin decided to establish GEV.[24] Aside from learning, these individuals also appeared as 'ambassadors' of China in introducing the public's commitment to environmental protection. For the first time, Chinese environmental activists made their foreign counterparts recognize that independent NGOs exist in the field of environmental protection in China, as opposed to being only the state's subordinate organs. In this way, more agencies and NGOs abroad become interested in initiating environmental programs in China and the local ENGOs benefited indirectly from this process.

In short, Beijing's ENGO community represents a contradictory feature: there exists a clear informal structure within and among ENGOs; at the same time, under the influence of international groups, local groups develop certain formal profiles. In order to explore the characteristics of the environmental movement in Beijing, network analysis is utilized in this case study.

Methodology

The Green Environmental Volunteers has been selected as a case study. The GEV is a typical influential grassroots group in Beijing's NGO community. This group has been selected for three reasons. First, the GEV is an influential national group that has strong mobilizing capacity among the public. Its permanent members reach around 200 through the years, while it has a variable number of volunteers that has reached more than 30,000 since it was established. As such, it is one of the largest ENGOs in Beijing. Second, the GEV has built rich contacts with organizations, environmental activists, and INGOs in Beijing. As mentioned above, it has been a leader in several environmental campaigns since 2000, and thus has a very influential position among the Beijing environmental movement. Third, the GEV is widely recognized by international NGOs and foundations, which provide ample funding to the group. Since its establishment, the GEV has received funds from the Global Environment Fund, the Ford Foundation, the Royal Embassy of the Netherlands, Ecologia (USA), International River Net, and Conservation International. In this way, the GEV proves to be a group with social leverage among ENGOs in Beijing.

The field work for this study was conducted from April 2004 to March 2005. The methodology applied to collect data was participatory observation and face-to-face interviews with a semi-structured questionnaire consisting of open questions. Being a member of the research staff from the NGO Research Center at Tsinghua University, the author was accepted to conduct research with this group. She participated in numerous ENGO gatherings and campaigns. These participatory experiences provided a rich source of information on the functioning of these movement networks.[25] Interviews were conducted on leaders and founders of GEV, its staff and members. By a snowball method, individuals in personal networks were selected and introduced one by one. In this way, information on the functioning of the personal network could be gathered. News reports, documents and existing literature, and interviews that have been published were also collected and used.

Organizational development of GEV

Formality of organization

Originally, the GEV was a recreational association. During a number of years, the GEV's mainstream activities focused on bird watching and raising environmental consciousness among city-dwellers. Its campaign changed frequently over the years. Issues such as the protection of resorts, poverty reduction, and dam protests were covered, and since 2002 the main attention is focused on protecting rivers and protesting against dam construction.[26]

The GEV's institutions are especially loosely built, and are characterized as an informal social organization. During the first five years, the GEV had only three full-time staff members, Ms Wang (leader), Ms Jin (leader) and Ms Zhang (administrator). In 2004, a capable figure, Mr Zhao, joined and took charge of daily management. Ms Jin, however, left around that time.[27] The GEV does not have a permanent workplace. Depending on the funding and the projects it handles, it rents offices. There was no formal internal decision-making mechanism before 2003. Decisions were usually taken by the two leaders, and implemented in a top-down fashion. After Ms Jin left, Ms Wang made decisions by herself, which were disseminated by telephone or email. Besides this, the GEV does not have clearly established institutions. It has no rules in operating projects, nor regulations in operating the GEV's income. The internal management did not allow the involvement of its members, who mostly participate only in activities of GEV.

Regarding the management of members, no systematic institutions are established to regulate membership. Individuals are accepted as members once they participate in activities of the GEV. Members' presence in the activities is not compulsory; attendance is voluntary. Members also have to cover their own costs in excursions and during activities. Their behavior is 'ruled' by trust from the leaders and individual self-consciousness of members. The communication between members and leaders is conducted through email or phone, which has become the most important method of binding members together.

In summary, the GEV has the organizational structure of a typical informal social organization. Over the years, the GEV's way of operation has been widely recognized and valued among its members. It is hard to see any institutional reforms in the GEV in the near future.

Leader's personal network

The informal network of the GEV's leader and founder, Ms Wang, played a crucial role in the management and operation of the group. Her personal network contributed greatly to this organization's growth and impact.

Being an active journalist and environmentalist, Ms Wang has an extensive personal network which is primarily constituted by pre-existing personal connections. The network members who are still active in Ms Wang's environmental activism can be divided into two main parts. One is made up of activists from the Student's Movement in 1989.[28] Introduced by friends, Ms Wang became familiar with a small number of civic activists, for instance Dai Qing.[29] By meeting and discussing with these individuals, Ms Wang learned useful tactics that could be used in raising environmental campaigns to affect the current environmental policies. The second part of the network is composed primarily of contacts she accumulated in her news reporting, and listeners to the radio program that Ms Wang chaired in Beijing Radio Station since 1988, *Environmental Protection Focus* (*Huanbao Rexian*). The listeners became loyal followers of Ms Wang in her environmental protection activities, and many of them became members of GEV.

The composition of Ms Wang's personal network is shown in Table 4.1 by occupation, income and age levels of members of the GEV.[30] The majority of members have an average income of 2000–5000 RMB (app. US$300.0-US$750). The average monthly salary of employed citizens in Beijing in 2004 was 2362 RMB (app. US$354.3).[31] The core part of Ms Wang's personal network is composed of environmentalists and journalists. Journalists are significantly involved in activities of the GEV and many of them participate in the 'journalist salon' and 'bird watching'—fixed programs of the GEV.[32] NGO members form another major part of Ms Wang's personal ties. Leaders of several important ENGOs are closely connected to Ms Wang, such as those of the Beijing Brooks Education Center, the Friends of Nature, the Green Island, and the Institute of Environment and Development. A smaller part of Ms Wang's individual network is composed of governmental officials and the members of

Table 4.1 Characteristics of members of GEV

Characteristics		*Percentage of members*
Gender	Male	40
	Female	60
Age	Under 25 years old	25
	25–35	20
	36–45	15
	>45	40
Monthly income (RMB)	<2000	10
	2000–5000	75
	5000–10,000	10
	>10,000	5
Occupation	Government	10
	Journalists/media	30
	Teachers in primary and secondary school	10
	University staff	10
	Other NGOs	30
	Freelance	5
	Other occupation	5

the scientific community. As a last characteristic, women are more strongly represented in Ms Wang's personal network than men, which is common in other environmental movement organizations as well. According to one NGO leader, women represent some of the best virtues in Chinese society: they are determined, patient and affective; and are not just pursuing fame or money.[33]

Most connections within Ms Wang's network, and also most members of the GEV, are over the age of 45. Thus the largest cohort experienced hardship in their youth, when China was a developing country. They witnessed various political movements of the Chinese Communist Party (CCP) that disturbed many lives, the most well known being the Cultural Revolution. These older generations also gained a large sense of social responsibility towards collective interests (Liu, 1992). At the other end of the age spectrum, we see a significant cohort of young people. Among them, people born in the 1970s and early 1980s also compose a large group. Their participation was caused by their share of sense of social responsibility with the older generation. Taught by moral education that was compelled by the ruling party in their childhood (e.g. Yuan & Shen, 1998), this group of young people is deeply influenced by social ideals and collectiveness, as emphasized by the CCP.

The construction of Ms Wang's personal connections with environmentalists is based on sharing a particular kind of environmentalism, which is a combination of social responsibilities and environmental care. This sense of environmentalism is rooted in a so-called elitism that commits to improving social equity and realizing one's ideals to promote social reform. These feelings are particularly prevalent in Ms Wang's personal network, composed of officials, scholars, journalists, and NGO professionals, who have resources or are in position to control the flow of resources. This group of people—usually middle aged—has a higher likelihood of believing in Confucian principles (Shi, 2000). With respect to the positions they occupy, they represent social elites in Chinese society (Li, C., 1996). Under the influence of the principles of Confucianism, these individuals believe it is an obligation for them to articulate the public's interest in environmental protection, although such deeds require tremendous courage and strength under the authoritarian rule of the CCP. A kind of 'environmental elitism' was formed. The life story of a scientist—Huang Wanli—is perceived as spiritual encouragement. Being an honest expert, he strongly objected to the construction of the Three Gorges Dam and did not follow the scientific projects of the CCP. Most of the time he was one of the few who faced the authorities. To the members of the GEV, this represented the strength and persistence that elites should have to protect an endangered environment.[34]

The idea of promoting the environment is equally prominent as a deep commitment. Generally speaking, this idea can be seen as one way of realizing one's social ideal for a responsible individual. This sense of solidarity grows naturally along with the rising of environmental awareness among the public. As described by the leader of the Green Island, part of Ms Wang's personal network:

> 'I like to have commitments in the field of environment: it keeps me moving ahead—to organize more activities, and to work more for protecting the environment … In the interactions with volunteers, I feel very much encouraged and motivated. I am doing this work with love. I believe Ms Wang shares this with me too.'

Ms Wang's identity tends to be a personal identity, which is not so much shaped by the GEV, but rather is formed by past and present interactions with individuals of her personal network. Through her, such identity is strongly strengthened and disseminated via the GEV network to become a collective identity of this organization. Viewed by the core members of her personal network, there exists a large homogeneity among GEV members. Although these individuals are part of the GEV, their identity was more prominently formed in their connection with Ms Wang's personal network. Thus Ms Wang's personal network has become powerful in organizing individual participants.

As to the interaction between Ms Wang and members of her personal network, equity and voluntary principles are the main principles used in the construction and maintenance of personal relationships. Usually, benefits and incentives are available for both counterparts. Members can withdraw certain resources that interest them from GEV through Ms Wang. For instance, journalists could get updated news on environmental protection or campaigns in the 'journalist salon' of GEV. As to the stability of organizational links, it is found that GEV is not immune to the central position and role of Ms Wang, who a founder and charismatic leader. However, this dominant position does not strongly affect the agenda of GEV, because of GEV's broad area of interests. This organization has rather diverse focuses, such as environmental education, protection of wild birds and animals, and desertification problems in northern China. Recently, it turned to the area of protecting rivers. In this way, GEV could relatively easily build organizational links with ENGOs that share its interests and goals. And its leader does not raise controversies on such issues. But this situation is likely to change when the organization become more focused on its area of activities. The influence of the leader's personal network will affect its focus and development path, which can lead to fragmentation of the organization. Yet as long as GEV continues to have diverse areas of activity, it will not be affected by its leader's personal influence.

The campaign against hydro-electric power projects on the Nu River

The Nu River is one of the last two large rivers that have not yet been dammed in China. Originating in the Qinghai–Tibetan plateau, it flows through Yunnan into Myanmar, where it is named the Salween River. The river passes through spectacular gorges, and its waters wash the habitats of many plant and animal species. As one of the Three Parallel Rivers—the Nu, the Jinsha (Yangtze) and the Lancang (Mekong)—its watershed is part of a World Heritage Site, which is the epicenter of Chinese biodiversity containing virgin forests, 6000 plant species and 79 rare or endangered animal species (Map 4.1).

In July 2003, a report on 'Planning of Electricity Exploration in Middle and Lower Stream of Nu River', initiated by the National Investigation Institute and the Kunming Investigation Institute, was accepted by NDRC. This report detailed the plan for constructing 13 hydro-electric power stations in the Nu River with the objective of producing 3.64 million kilowatts of electricity to feed China's tremendous appetite for power, which is even more than the Three Gorges Dam. The project facilitates China to address energy shortages while bringing jobs to poor residents of southwest China and revenues to the provincial and local governments. However, the various dams would seriously affect the natural flow of the river, with significant environmental and social effects—but these consequences played a minor role in the report.

One month after the launch of the report, a high-level meeting was organized by NDRC to discuss the report on 'Planning of Electricity Exploration in the Middle and Lower Stream of Nu River'. Although MEP expressed its reservations and disagreement during this meeting, the project proposal was accepted by the meeting and delivered to the State Council. The proposal planned to start dam construction by the end of 2003, but it became an object of environmental and social controversy.

Around this conflict, two sides can be identified. MEP and a coalition of environmental movement organizations formed the core of the opposition against the plan. In September and October 2003, MEP started to organize expert meetings and tried to mobilize public opinion on the negative consequences

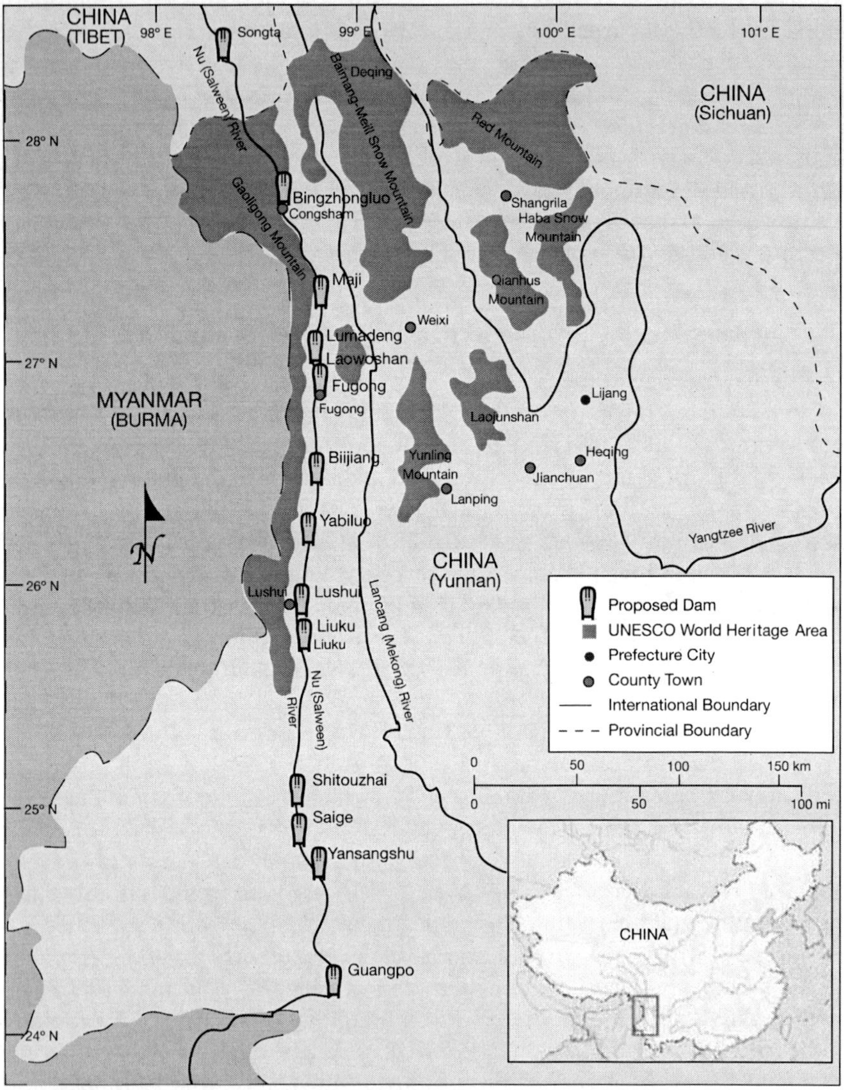

Map 4.1 The Nu River and the proposed dams

Source: International Rivers (http://internationalrivers.org/en/china/nu-salween-river/nujiang-map)

of the project proposal, but this did not make much difference.[35] The campaign of Chinese ENGOs was initiated in August 2003. According to them, the construction of dams would forcibly displace 50,000 people, indirectly affect the livelihoods of millions living downstream in China, Burma and Thailand, and negatively affect the flora and fauna in the surrounding areas.

The core of the side supporting the dams consists of NDRC in the central government and the local authorities in Yunnan Province. NDRC claimed that the amount of electricity resulting from the dam construction would solve the nation's current energy crisis. The local government in the Nujiang area is not strong, is not rich enough to initiate such a construction project, and was very pleased to obtain such a large investment project from the central government.[36] Ecological and social effects of this construction were of minor concern to the provincial and local governments in Yunnan.

After a first delay, the conflict culminated in the decision of premier Wen Jiabao in April 2004 to halt implementation of plans to build a 13-station hydro-electric dam on the Nu River. He called for careful consideration of major hydro-electric projects 'that have aroused a high level of concern in society, and with which the environmental protection side disagrees'.[37] The various activities of NGOs began to decline in August 2004. However, policy discussions on the dam construction project were resumed in October 2004, when the pro-dam authorities tried again to put the project on the political agenda. Although the plan was poised to move ahead, the project developer, China Huadian, was reported to begin preparations in February 2008. It is still unclear whether the project received final approval from Beijing. Yet, local government officials denied that approvals had been given and any construction has started.

Establishment of the social movement network

In this campaign, GEV played a particularly important role in constructing a social movement network around this campaign, based to a large extent on the informal network of its leader. This movement network is constituted by activities organized in Beijing and Kunming.

Composition of local environmentalists

Kunming, the capital of Yunnan Province, where provincial government and a number of NGOs are situated, composed a part of the movement network that was facilitated by GEV. Here, Green Watershed, a grassroots NGO, and scholars and experts from Yunnan University (located in Kunming) who are familiar with this issue, are the main actors.

In the local ENGO community, no ally among actors of civil society could be identified, owing to the repressive attitude of the provincial government on this issue. Mass media in this city are strictly administered by the authorities, which prevented them from reporting articles that object to governmental policies. Besides, local journalists show little concern with the development of ENGOs,[38] so they are hardly ever included in GEV's movement network.

Regarding the local ENGO community, although there are many ENGOs located here, no collaboration existed in the ENGO community before the GEV campaign.[39] Very few ENGOs responded to this issue, and the more influential the ENGO, the more likely that it avoided this issue, fearing endangerment of its status.[40] The Center for Biodiversity and Indigenous Knowledge, a well funded group that is committed to promoting local participatory governance and biodiversity protection, attempted to produce an internal report directed at the provincial government.[41] Yet, considering the political sensitiveness of the issue, they did not realize this plan. Green Watershed strongly objected to the project and actively provided news to the public to build up protest. Being a grassroots group that has relatively little contact with local officials, Green Watershed is weak in exerting pressure on politicians. As an alternative, they turned to GEV and Beijing's ENGO community to strengthen their influence.[42] Although a number of INGOs are also located in the city, they did not organize any form of objection or protest. Fearing that expressing their objection might endanger their status and their permit to work in China, these influential INGOs took a conservative stance on this issue.[43] As to local scholars who expressed opposition to the policy, under pressure from the local government, they were prevented from expressing opinions that were critical of dam construction.

In general, the local ENGOs that protested against the dams were relatively weak. They could hardly exert pressure on Yunnan Provincial Government on this issue.

Structure of movement network

By utilizing Ms Wang's informal network, GEV played an important role in this campaign by liaising on an individual level. As a result of these links, a movement network was formed around the campaign, composed of ENGOs, MEP, media and scientists.

The network of ENGOs in Beijing appears to be the basis for building a movement network. Concerning this campaign, GEV proved to have a strong networking capacity through Ms Wang's personal network with the leaders of several ENGOs. It could occupy such a central and coordinating position in the campaign due to Ms Wang's *guanxi*. Educational NGOs (Friends of Nature, Beijing Brooks Education Center, Global Village Beijing), a research institute (Institute of Environment and Development) and students' environmental groups cooperated in the movement network that GEV initiated around the Nu River dam campaign.

GEV's link with MEP was also of major relevance for the campaign and for the final outcome. In the policy debate around the Nu River dams, the NDRC was the main actor initiating this project, and was also responsible for organizing the environmental assessment. MEP, of the same rank as NDRC, was left with limited authority in influencing the initial policy proposal, the subsequent policy process and the organization of the environmental assessment. MEP's demand for a strict environmental assessment of the project plan by neutral institutions was not followed.[44] In order to gain leverage on this project, a minority of officials in MEP decided to draw support from environmental movement organizations. The collaboration was based mainly on Ms Wang's close connections with Mr Mu, previous vice chancellor of the supervision department of MEP. Mr Mu's clearly positive attitude toward ENGOs was known and silently approved by a small group of top leaders in MEP.[45] During the policy process and the campaign, Mr Mu provided Ms Wang and GEV with updated information, both on substantial environmental information and on the development of the political debate. Thus information and news became a very prominent resource in Ms Wang's personal network. And the capacity to provide information on policy and political dynamics founded the central position of Ms Wang and her GEV in this movement.

Mass media form an important counterpart of the environmental movement, especially in gaining influence. Media coverage can raise the profile of ENGOs; gain public support for the anti-dam campaign; and bring a form of recognition from the state as well (as it does not suppress this media coverage). With Ms Wang's tight connections to journalists at key journals and newspapers, GEV could build close links with the media, develop public opinion, and build consensus.[46] Constructing public consensus was one of the main strategies in the anti-dam campaign.

Scientists are generally valued highly by Chinese policy-makers (Sullivan, 1994). In the past, experts and scholars used to be controlled tightly by their *danwei* (institute or research center), and thus controlled by the Party State when expressing their opinions and ideas. Along with the decline of *danwei* in Chinese economic reform, this group of individuals become relatively isolated actors in society, and began to bravely express objections to the Party's policies. As a journalist, Ms Zhang's own experiences witness the changing attitude of Chinese individual experts:

> 'In the middle of the 1990s, I conducted reports on science and development, which provided me many opportunities to interact with scientists. They were a lot more cautious when expressing their ideas, comparing to the attitude of scientists nowadays. Whether they are in their 30s or in their 70s, the scientists I contacted are willing to object to government. They are quite courageous and determined in doing so in recent times.'[47]

In this controversy, science developed into a key issue, as MEP accused the environmental assessment on dam building to be subjective and one-sided. Opinions of scientists and scholars became important arguments for MEP in policy debates and for the movement coalition in campaigns and media. Through Ms Wang's informal network, GEV was capable of building networks with experts and scholars in Beijing, although its role in building contacts with local scientists was not large. Such contacts were built with Yunnan scientists through Green Watershed as a local force. Because local scientists are more familiar with the specific local situation, integrating local scientists in its network became a critical issue for the environmental movement.

As a result, in this campaign, the movement network was most active in the capital, Beijing. It adopted two strategies to influence policy- and decision-making. First, it mobilized public opinion and tried to build a collective consensus against dam construction, not unfamiliar in Chinese decision-making processes (cf. Lieberthal & Lampton, 1992). As a result, a nationwide public consensus was mobilized. Second, influential collective activities were organized that had an impact on society and policy-makers.

Building collective consensus

Through its close links with MEP, various journalists and the media, the environmental movement against the Nu River dam was able to develop a strategy of regular publications in local and national journals, radio and television. Social influence was mobilized this way.

Two periods can be distinguished by examining the issues that were central in this strategy of building public opinion. From August 2003 to February 2004, the aim of publications and information dissemination via the media was to attract public attention to the controversy of dam building on the Nu River. More than 20 newspaper articles concerning the dam project and the controversy around it were published. These articles were published not only in general newspapers for the wider public, but also in the daily newspaper of the CCP, which has a major impact on policy-makers.[48] At the same time, a TV program was made and broadcast.[49]

In this period, consensus-building by GEV and its coalition was very much focused on ecological and environmental issues. MEP's early opinion, 'protecting the river that has great ecological significance', formed the starting point and was continuously repeated in journal articles.[50] This argument became the major focus of the pro-construction side, which constantly insisted on judgments based on growth of the local economy and the increase in national energy, using the ecological assessment report to deny serious ecological effects. Rather than relying on MEP's judgment and impact, GEV increasingly called in experts, who used scientific, ecological, and environmental arguments. The expert resources were again widely disseminated via the media in order to build public opinion and consensus.[51]

The repertoire changed in the second stage, along with GEV's deeper understanding about the project. In February 2004, Beijing-based GEV organized a mission for journalists, Beijing's scholars, and environmentalists to investigate the local situation of the Nu River. During this trip, participants became impressed by the potential impact that the dam construction project could have on the rich biological and cultural diversity in the area. The social consequences were only then fully understood and 'experienced'. Twenty-two minorities and six kinds of religion coexist in this area, most of whom farm and herd in the isolated mountains above the river. The project would result in the relocation of 50,000 people, especially from the minorities. This excursion led to a large number of reports and articles, especially regarding the social conditions in the Nu River area and how these would be affected by the dams.[52] During this period, consensus-building shifted from mere ecological protection to include concern for the inhabitants.[53]

The previous consensus-building that focused mainly on ecological values was vulnerable to scientific debate and counteracted by arguments regarding the need for local economic development. But the new consensus was broadly accepted, and directed attention to social problems in the local area. Arguments of economic development by NDRC and local governmental authorities proved of no use in attacking this new consensus. This media strategy and consensus-building also worked effectively for the various environmental movement organizations, bringing them together and pushing the campaign to a new stage.

Formation of coordinated actions

In the latter period of the campaign, ENGOs collaborated effectively. They organized coordinated actions, although the Chinese context limited the forms of collective action that could be achieved. The purpose was to publicize the campaign constantly and produce a greater impact on society.

In Beijing, photo exhibitions were held in the universities. In March and April of 2004, for instance, nine photo exhibitions were held in Beijing's universities, a supermarket, an office building and a post office. Student environmental associations were contacted to help to arrange the time and venue at their campus.

In order to generate greater influence, an interactive website called 'Nu River Sentiment' was established. This website facilitated exchanging information and communication among movement actors—the progress of the campaign was described, photos and articles on dam construction were published, and scientific information was provided. Various ENGOs joined in building this website, declared their support to the campaign, edited the content, and provided technical support.

In the campaign, ENGOs also tried to influence the development project by participating in institutional channels. Friends of Nature, an influential NGO that is relatively strong in accessing political institutions, was included in the network. Through its influence in political and policy institutions, it helped to introduce scientific reports made by GEV-related experts at various points in the policy-making process. Two major studies, 'Protecting Nu River, Stop Water and Hydro-electric Development' and 'Grouping Rivers, Coordinating Ecological Protection and Economic Development', were brought into the National Political Consultative Conference in March 2004 and became part of the political debate there. Personal ties between policy-makers, politicians, and leaders of the Friends of Nature and other environmental movement organizations also proved to be instrumental.

In addition, actions were organized in order to have international influence through international conferences. For instance, at an international conference, 'World and People along Watershed', held in Thailand in December 2003, representatives of Chinese ENGOs, including GEV and Green Watershed, initiated a signing session to protest against the dam construction. As a result, 80 NGOs from Thailand and Laos that are situated in the lower streams of the Nu River collectively wrote to the Chinese Government, inquiring about decision-making on this project without consulting the downstream countries. This letter was sent to the Chinese embassy in Thailand.

Conclusion

The campaign of environmental organizations to stop the hydro-electric project on the Nu River can be interpreted as a milestone in the Chinese environmental movement. For the first time, ENGOs were successful in, at least temporarily, stopping a major infrastructure project that was endorsed so strongly by the local government and NDRC.

Although this case study saw the emergence of an effective movement network among environmentalists, scholars, officials, and ENGOs, it was the informal network of the leaders of the environmental organizations that was crucial in organizing the campaign and mobilizing support. By distinguishing informal networks, it became apparent that informal networks (around one individual) could explain the organization of the campaign and the successful use of various resources.

Considering the development of GEV, its foundation for building an informal network is rooted in local culture, rather than its organizational capacity. Being among the earliest NGOs to be established, GEV represents characteristics of an influential national group, although it preserves its informal organizational form. This tactic has been used for years after it was established, with no sign that any institutional reform will take place within the coming years. It can be concluded that it is the local culture that fosters the practice of personal networks. In this way, a wide range of actors are organized, with a plurality of movement actors being mobilized in the movement network: not only organizations, but also individual actors such as scientists.

What is more, this network is not limited to Beijing, but extends to local environmental activists and scholars. The broad movement network is constructed of individual ties, based on sharing the collective identity that Ms Wang advocated through the movement. These connections were built relatively easily and can be expanded in geographical scope throughout the country. And so, through a personal network, a kind of 'environmental elitism' combined with environmental affections is achieved and communicated among counterparts. In this way, the content of an informal network determined, to a significant extent, the nature and the identity of the movement network. Through this, coordinated actions were realized. In terms of Diani's (2003) distinction between 'brokers' and 'leaders' in movement networks, GEV and Ms Wang clearly acted as a leader rather than a broker.

In highlighting the innovations in the Nu Dam campaign, it is interesting to compare this campaign with the environmental campaign on the Three Gorges Dam project, which took place in the early 1990s. In both campaigns, raising public opinion and consensus-building was a main strategy, and scientists and policy-makers were equally important actors. However, the differences are even more striking than these similarities, both in the campaign and in the final outcome. The different political and social context of the Three Gorges Dam campaign did not allow the emergence of an independent network of environmental movement organizations and individuals; instead, the opposition remained restricted to individuals, rather than organizations or a movement. Also, the alliance with the media makes this campaign different from the one around the Three Georges Dam, when the media were still strongly controlled by the state. Furthermore, the fact that the movement network incorporates MEP marks another feature of the development of environmentalism. In comparison, during the Three Gorges Dam project controversy, the political totalitarian system lacked any checks in and outside the political bureaucracy. So the dominant large-scale bureaucracies, the Ministry of Water Resources and the Yangtze Valley Planning Office, could fully determine the outcome.

5 Environmental activism in Xiangfan, Hubei Province

Introduction

The local government and local civil society

Xiangfan city, the second largest city in Hubei Province, is located in the middle part of Han River—the longest branch of Chang River in the middle of China. It is situated at the confluence of the Hanshui, Tanghe, and Baihe rivers in northwest Hubei Province. Xiangfan is composed of two originally separated cities, Xiangfan and Fancheng, with a total population of 5.8 million, located in 12 administrative areas covering 19,000 square kilometers.[1] The average GDP of Xiangfan was 12,310 RMB (approximately US$1518) in 2005, which is a medium–low level of economic development compared with the rest of the country.

Xiangfan is undergoing an economic transition, with the goal of developing towards an industrialized city. The metropolis is one of the industrial and agricultural bases for the entire country. Manufacturing accounts for 46.4 per cent of the total economy, which makes the largest proportion of the local economy.[2] Mainly represented by the car, spinning, medical and chemical industries, and new high-technology industries, manufacturing industry is strongly promoted as the core of its total economy.[3] Besides this, Xiangfan has favorable natural conditions for agriculture: it has a pleasant climate and a rich ecological environment, and Xiangfan was developed as one of the most important grain-production bases by the central government under the planned economy. With 4160 square kilometers of farmland, it is now one of the 10 biggest summer grain production bases across the country (Xiangfan Publicity Dept, 2003).

In the process of economic development, Xiangfan's market economy develops at a slow pace. Under the strong control of central and local authorities, the public-owned economy has been the dominating model since the early years after the Communist Government was founded in 1949. The economic reform did not, however, successfully establish a market system. The state-owned economy still functions as it did under the planned policies, rather than through market mechanisms. The state-owned economy also makes up a large part of the total economy, larger than the private economy. Its persistence further obstructs the establishment of a private economy,[4] which grows slowly. Only in 2003, a growth of 3.6 per cent was realized in private economy which appears to be a highest percentage realized in years. As the latest statistics show, the proportion of private economy in wealth creation reached 39.6 per cent in 2004, which is still 18.2 per cent lower than the country average.[5]

Affected by its economic structure, Xiangfan's political development appears to be that of a monopolist that controls resources and power. This can be shown in two aspects. First, the government does not release freely useful environmental information, and the public cannot access these data. Second, political organizations possess the authority to administrate public affairs and are widely recognized by the public. Other actors, including social organizations, lack legitimacy, from the point of view of the residents, when conducting their activities without utilizing administrative channels of policy implementation.

More specifically, political institutions administer the development of civic organizations, which do not have complete autonomy. According to 'Regulation on the Management of Social Organizations', prescribed in 1989, government agencies can administer forces of civil society in a wide range of issues, ranging from registration through daily operation to internal organization. Due to this situation, very few environmental NGOs (ENGOs) exist in Xiangfan. Green Han River (GH) appears to be the only independent environmental group in the city.

Beside the repressive control, local government agencies have limited knowledge of the representatives of civil society due to the slow emergence of these organizations. Their attitude toward NGOs is rather restrictive. In August 2003, the GH invited Beijing's and Tianjin's NGOs to Xiangfan and organized meetings with the city government in order to make clear what losses Xiangfan would face by the implementation of the South–North River Diversion Project scheme.[6] Fifteen ENGO professionals from 11 ENGOs in Beijing and Tianjin were invited to Xiangfan to investigate the local South–North River Diversion construction. The Xiangfan Government feared that the dissident opinions of local citizens would be taken as the Government's attitude, and would be disseminated by NGO activists from Beijing. The Spiritual and Civilization Construction Office (SCCO)—a government agency that cooperated with GH—took over the organization of this tour from GH. The purpose of the original tour was then changed to a large extent. Political officials took charge of the tour, and environmentalists lost the freedom to get in touch with local environmentalists. Unwillingly arranged in this situation, the NGO activists felt that GH had not been recognized as an autonomous actor by the political authorities.

Although the political system controlled the actors in society to a large degree, the mass media appeared to be among the first group of actors that gained a certain degree of independence. Local media began to differentiate from the state authority to become independent market actors. Competition exists among a number of local mass media agents. Two relatively autonomous local newspapers are active in Xiangfan: *Xiangyang Daily* and *Xiangfan Evening News*. *Chutian City News* has a specific focus on Xiangfan, which also makes it a competitor of the local newspapers, although it is located in Wuhan, the capital of Hubei Province. The mass media have thus gained a certain degree of autonomy.[7]

Local culture of guanxi *operation*

Xiangfan has a significant history in China, which leaves the area a rich traditional cultural heritage. The preservation of *guanxi* practice is related to the cultural heritage of Xiangfan. Built around 2800 years ago, Xiangyang, the old part of the city of Xiangfan, has been important in Chinese history (Wen, X., 2006). It used to be a crucial transportation site on the route from south to north, and thus comprises one of the cultural centers in China. Furthermore, a number of famous people were born in this city, such as Zhuge Liang, one of the most well known politicians and strategists in Chinese history who helped Liu Bei establish and consolidate his rule during the Three Kingdoms Period (the third century AD). It was also the hometown of a number of great artists and authors.

The historical past of Xiangfan has shaped its contemporary society. Xiangfan city has preserved values and principles of traditional Chinese societies relatively well (Wen, X., 2006). The crucial role of *guanxi* in social life became prevalent in Xiangfan. Particularly, the function of building personal relationships and utilizing them for accessing resources is widely recognized in this city, and is adopted by ordinary citizens in their daily life.

Methodology

The field work for this study was conducted twice, in December 2003 and September 2005, respectively. Triangulation was applied in this study to collect data, including semi-structured face-to-face interviews, documentary research and a participatory approach. Being a member of the research staff from the NGO Research Center at Tsinghua University, the author joined the project initiating environmental governance in Wushan County by Green Cross, which started in the end of 2003. She was further allowed to conduct research with GH. Around 40 interviews were conducted, including leaders and members of GH, local officials and ENGO staff in Beijing, with whom the organization mainly interacted. In this way, information on the functioning of personal networks was gathered. Besides this, news reports, documents and existing literature, and published interviews were also collected and analyzed.

Organizational structure of Green Han River

Green Han River, a grassroots ENGO, was founded in 2002. Its goal is to raise the public's environmental awareness, and the main area of activity is environmental education. Although the GH has 122 individual and 41 collective members,[8] it relies heavily on its three full-time staff.

Since being established, the GH has been autonomous from the city government, mostly because, unlike most local social organizations, it does not rely on government subsidies. The GH primarily relies on itself for funding. The GH's income is mainly composed of three parts: aid from international foundations, government subsidies, and membership fees (Figure 5.1).[9] Financial support from international organizations accounts for the majority of its income since its establishment in 2002. The major international sponsors are the Ford Foundation, the Global Grant Foundation, the Canada Civil Society Fund, and the World Bank. A small amount of income also came from the government to subsidize particular projects in which government actors participated. For instance, in the project 'Ecological Education in Rural Areas', implemented in 2004, the Bureau of Technology subsidized 4000 RMB (approximately US $600) to cover the operational costs of the project. Only a small amount of the income came from membership fees in the past three years. The annual revenue was around 70,000 RMB (US$10,500).

Members of the GH comprise an important part of this organization (Tables 5.1 and 5.2).[10] Two groups of members can be identified among them. Group A is organized through rules and institutions; while group B is mobilized mainly through the leaders' personal network.

Formal mechanism of group A

Members in group A are mostly young members under 45 years old, primarily recruited as participants in concrete activities and mass campaigns (Table 5.1).

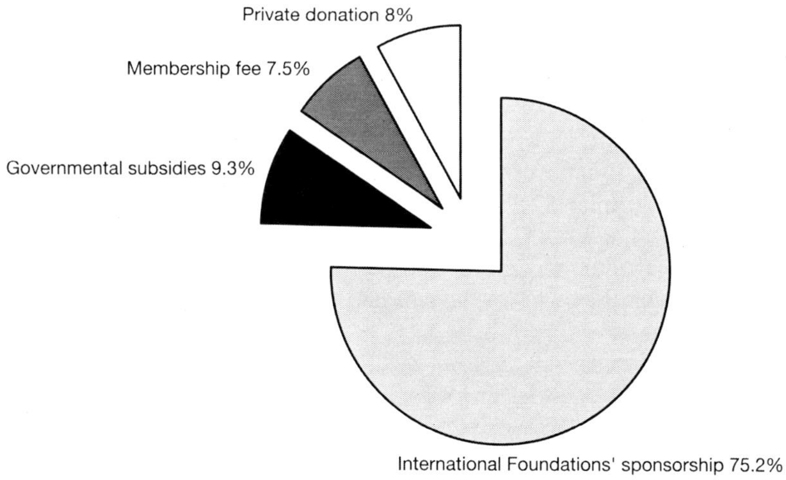

Figure 5.1 Composition of Green Han River's total income (2002–05)

Table 5.1 GH membership composition (group A and B, total members = 122)

Age	*Percentage*
<25	1.6
26–35	7.3
35–45	43.4
46–55	21.4
>56	26.3

Source: GH member's registration book.

Table 5.2 Occupations of GH members (group A and B, total members = 122)

Occupation	*Percentage*
School teacher	14.7
Journalists	13.4
Governmental units	39.1
Xiangfan City People's Political Consultative Congress and Xiangfan City People's Congress*	13.1
Company	13.1
Others	6.6

Notes: *Chinese People's Political Consultative Conference and Xiangfan City People's Congress. CPPCC is a political advisory body in China. The members' term of office lasts five years. GH members include those who are serving as present members of CPPCC, but also retired members. Xiangfan City People's Congress is the legislative institution. At city level, members are elected indirectly from district level. One term of office is five years.

Young citizens in Xiangfan city apply eagerly to become members of the GH. They enjoy participating in activities that show their concern about the environment. Collective identity is built on two main dimensions. First, individual members have a relatively high level of environmental awareness. Such consciousness stems from love towards their homeland, especially when serious environmental degradation occurs.[11] As suggested in the literature on social movements, geographical boundaries are important in forming collective consciousness and identity (McCarthy & Wolfson, 1992). Being inhabitants of Xiangfan, these young people formed their environmentalism out of love for the place in which they live. In the words of a member:

> 'I tell you, we are saving you, Bai River/ to return you your beautiful smiles are our wishes/ let us love you all together/ a beautiful home needs love from every one.'[12]

The second dimension of collective identity shared among members is devotion to the GH, rooted in the culture of Communism. As a result of the strong control of the government, the values that the Chinese Communist Party (CCP) raised to regulate its members in the 1960s are still very influential among Xiangfan's inhabitants. It advocated that CCP members should commit themselves to the public good without considering self-interest, and serve the people 'wholeheartedly'. These virtues are believed to be beneficial to environmental protection. Expressed by one leader of GH, 'one should be ready to sacrifice oneself, because participating in environmental protection activities does not bring material profits.' As for recruitment, the GH follows strict procedures to select its candidates.[13] Its strictness in recruitment of members bears a certain similarity to the selection process of CCP members.[14] Members need to be individuals who have a good record at their work, and who show commitment to environmental protection. Also, as GH is an organization that recruits a large proportion of the social elite, it is very attractive to individual citizens, who aim to build personal relationships and utilize them for their own purposes.[15]

Being selected as a member of GH, they are expected to contribute to the organization without demanding benefits for themselves. They are required to obey the group's directions and be ready to make contributions to the group. Time and energy are the primary types of resources needed by the GH. The young people have to be present when GH needs them during the time they are off work. Sometimes they even have to provide their own possessions, such as cars, to facilitate GH's successful implementation of activities. On the other hand, members seldom ask for benefits in their own right, except from participation in the organization's policy-making. All in all, young members are required by GH to realize their obligations and demand little as members' benefits that GH cannot afford—except for opportunities for members to participate in excursions, no other rewards are available. Even so, the young people are attracted to GH and are relatively loyal to it. As they remark, they can represent the environmental interests of the public, and at the same time meet friends and have fun.

In order to be effective in organizing an environmental movement, the GH has also set up decision-making mechanisms, which are composed of vertical structures that follow democratic procedures (Figure 5.2). This system aims to suit the majority of GH members, who work in their spare time at the GH. All three founders, and now presidents, of GH—Ms Yun, Mr Li and Ms Ye, are full-time staff, and contribute a great deal of time to the organization's development.

Collective decision-making is a characteristic of the GH's system. According to the tasks and their frequency, three levels of meeting are held: the president's meeting, the council's meeting, and the members' meeting. The president's meeting is organized most often, and comprises a small circle of top officials, including the president and vice-president. It takes care of the administration and daily activities of the GH. Compared with the council's meeting, its authority on the group's development is relatively small. Yet the president's meeting does possess a degree of power over the GH's development. The council's meeting is held every six months, initiated by the group president. The council's meeting,

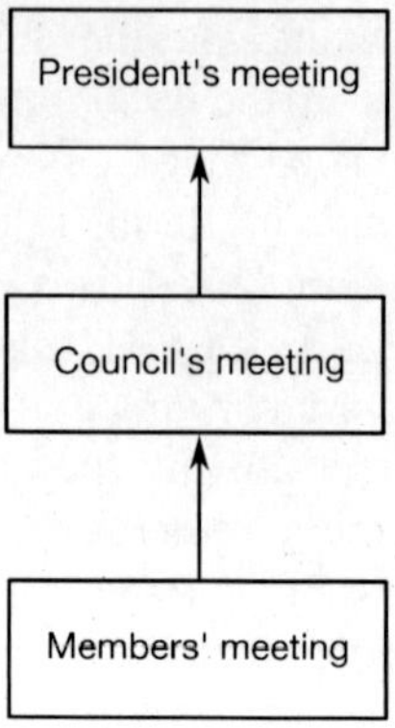

Figure 5.2 GH's internal organization for decision-making

members of which are elected by the members' meeting, has to report back to the members' meeting. It discusses strategic planning and activities of the GH in a longer-term perspective than the presidents' meeting.

As to the leadership system of the GH, a democratic system exists for selecting leaders. President and vice-president are selected from councilors. Councilors are elected by members of the GH during the members' meeting, which is held every four years. This system follows a democratic procedure, and is effective in making members realize their rights of management in the GH.[16]

In addition, specific mechanisms have been established to maintain members. For instance, the GH has built its own website in order to post information and reports of the GH.[17] Internal newsletters are produced and meetings held, through which information about the GH is disseminated among its adherents. Through these mechanisms, the GH builds a sense of belonging with its members, which makes it successful in mobilizing activists.[18]

In all, the organizational structure is established, with institutionalized mechanisms. Participants can thus realize participation in the management of the group. The GH has a high level of organizing capacity among the members of group A.

Informal mechanism of group B

Group B (47.7 per cent of GH members) mainly comprises people mobilized by Ms Yun, thanks to her personal charisma. Ms Yun, the leader of the GH, has a relatively broad personal network that is composed of local civil servants and members of Xiangfan Chinese People's Political Consultative Conference (XCPPCC) and Xiangfan City People's Congress (XCPC) (Table 5.2). These two groups account for a large proportion of the total number of members. These connections, partly through her work and partly through a shared residential area, form her personal network.

Through her work, Ms Yun has gained a rich experience in local politics. She worked for 20 years at the Office of Oversea Chinese Affairs in the Xiangfan Government before she retired. Furthermore, she was a member of the XCPC and XCPPCC, serving in each for one term.[19] Ms Yun developed close relations with her colleagues and co-members in these two institutions. Additionally, Ms Yun lives in the same area as the *danwei* compound of the institution of the XCPPCC and the XCPC.[20] Thus Ms Yun developed close ties with governmental individuals, whose homes are in the same residential area. These two factors enabled Ms Yun to build personal connections with civil servants and members of the XCPPCC and XCPC. Ms Yun's network is maintained in an informal way, and continued to exist even after she had retired. The longer they were maintained, the stronger personal relationships can become, because the personal ties that are developed bring a sense of familiarity and connection to the counterparts. For instance, as remarked by one of the neighbors, if you have been neighbors (colleagues) for so many years, it is very hard not to consider the familiarity that is developed with Ms Yun, and refuse doing a favour for her. Thus Ms Yun can successfully gain certain resources through people with whom she has close ties, resulting from the emotional dimension that their personal relationships bring.

Like group A, most members of group B are members of the CCP. They also share with group A the identity of making 'deep devotion' to the common development of the community. As most of them are of the retired political elite, they differ from group A in the specific ways in which they contribute to the public's interests. As some can still influence current political officials, even if they are in retirement, the goal to make 'deep devotion' to the common development of the community is shared among them.[21] For instance, as members of the XCPPCC, Ms Yun and her close connections Mr Li and Mr Ye[22] (also leaders of GH) are anxious to initiate public charity-related issues. As remarked by Mr Li: 'we are so used to working for issues concerning the public. We could not live without thinking of such issues.' So after acknowledging the Han River's deterioration, they were all anxious to do something to improve it. As Mr Li concluded, to improve water protection is 'the very right thing we should conduct'. Because such issues are so urgent, they should 'respond as quickly as possible'.[23] Ms Yun realized these individual wishes to improve the Han River's water quality and successfully utilized her personal network to mobilize interested individuals and construct solidarity to work for the environment. According to a member of GH, 'she worked like a flag that weaves us to progress in protecting the environment.'[24] In the ideas expressed by Ms Yun, values advocated by the CCP can be seen, such as self-sacrifice for the collective. She was thus praised and presented by the head of the city's Environmental Protection Bureau (EPB) as a model to the EPB's staff.[25]

Structure of organizational links

In order to initiate environmental reform, the GH had to build organizational links in order to access resources and mobilize coordinated actions. 'We need all kinds of resources, no matter [whether] non-material resources or material ones', as remarked by Ms Yun.[26] A dynamic and strategic use of personal networks is recognized in the interactions between GH and other actors. In addition, organizational networks are also built with a small number of collective actors.

Links with government agencies through personal networks

Under Xiangfan's socio-political conditions, government actors comprise the majority of partners. As analyzed in the Introduction, due to the dominance of political authorities, the government's involvement in environmental reform appears to be a necessity for any social forces to implement their activities successfully. Organizational links with government agencies are one of the most prominent parts of the networking activities of GH.

The GH collaborated with other political actors in order to access resources and influence in environmental reforms. In order to utilize political authority to a large extent, the governmental units chosen for cooperation were those responsible for the specific activity the GH was organizing. A great number of government agencies were contacted to implement the GH's activities, such as the Bureau of Water Management, the Bureau of Forestry, the Bureau of Education, etc. Among these, three constitute the major and lasting partners in the GH's environmental reform activities: Xiangfan City Bureau of Civil Affairs (XCBCA), the XCPPCC, and the local EPB. Personal networks were utilized when building these organizational links.

First of all, the GH had to seek resources to ensure its status as a social organization. It successfully built organizational links with the XCBCA, and secured its political status. The construction of these organizational links depended strongly on the personal ties between Ms Yun and the head of the XCBCA. Because these two people had been familiar with each other for a long time, they shared a high level of intimacy. This relationship was used as a resource to get the XCBCA to do favors for the GH. The XCBCA permitted the GH to be established without much complication, and agreed to act as the responsible and supervising institution for the GH, primarily as a favor to Ms Yun.[27] Only three months were spent in completing the legalization process, because 'he would have to give me *mianzi* when I applied to register GH', according to Ms Yun.[28]

What is more, the GH was also promoted as a candidate for prizes of distinguished social organizations. In 2004, the GH was promoted by the XCBCA as winner of the prize of 'model social organization', which was promulgated by the Provincial Bureau of Civil Affairs. This honor ensured the GH a secured political status in Xiangfan's political arena, without it being suspected of opposing local authorities, thanks to its political innocence. A relatively broad space was available for the GH to launch its activities in a conservative atmosphere like Xiangfan. Furthermore, annual investigations into GH's management were simplified. In recent years, because of the GH's politically innocent behavior, it gained more trust from the XCBCA and no annual checks were conducted at all.[29]

Second, the XCPPCC was another main actor with whom the GH built an organizational link. As a political advisory body, this organization works actively to articulate individuals' interests, and as an autonomous actor, it supervises the administrative organs.[30]

XCPPCC is an active actor in representing local environmental interests. It works independently in conducting environmental investigations to acknowledge citizens' interests, and then compiles reports to supervise top political leaders. In promoting local environmental governance, the XCPPCC collaborates closely with local EPBs to force local polluting companies to take measures in wastewater control.[31] Their reports are highly valued by the party leaders and policy-makers. By closely working with mass media, the XCPPCC also supervises environmental pollution by building public consensus, or setting up public hearings for citizens to discuss policies with officials.

The GH successfully builds collaboration with the XCPPCC. This organizational linkage was built through Ms Yun's *guanxi* network by working with its publicity unit, through her contacts with Ms Qiu, Director of the publicity unit of the XCPPCC. As neighbors in the same residential compound, the two women were very familiar with each other. Since both institutions they work for have similar goals, collaborating was a natural step for them. The fact that close personal relationships existed facilitated the construction of such linkages. In the two institutions' interaction, the GH's work provides material for the publicity unit of the XCPPCC: GH often conducts investigation on the Xiangfan's rivers' water quality, which can be presented as a live example for TV programs to raise the public's environmental awareness.[32] From the point of view of the GH, through collaborating with the XCPPCC, it had access to confidential information from the local water-quality test center on water-quality monitoring. Thus the two units worked closely in mobilizing the public consensus.[33]

Third, the Xiangfan EPB constitutes another partner of the GH. It lacks the capacity to govern the environment, due to its dependence on the city government and its low enforcement capacities. As the main agency of local environmental governance, the EPB is weak in initiating and mobilizing coordinated actions to introduce environmental reforms.

Administered by the city government, the Xiangfan EPB relies on limited revenue. Its income level greatly impeded its efficiency of environmental enforcement. In 2000, local EPBs' total income counted for 1.21 per cent of local revenue. The amount has reduced in recent years, and accounted for no more than 1 per cent of local revenue in 2005.[34] This is far below the requirements of the Ministry of Environmental Protection (MEP), which demand that the EPB should be financed by 3 per cent of the total local revenue. The budget that the EPB had could only afford a team of 22 staff members, while the supervising team that subordinates the EPB does not have a stable salary. Their incomes come mainly from a prescribed percentage of payment, which does not completely cover their salary. What is more, the EPB's budget constrains the department of pollution supervision and scientific research, which therefore lacks the facilities to carry out its tasks.[35]

The lack of autonomy of the EPB results in the EPB having incomplete authority in environmental governance. Regulatory measures, including environmental impact assessments and 'three synchronizations', cannot be introduced and enforced strictly. It is estimated that up to 70 per cent of all projects constructed have not made an environmental impact assessment (EIA).[36] Strong and influential local government agencies such as economic commissions, construction commissions, and industrial and commercial authorities, which constitute powerful units in the decision-making process, do not respect EIA procedures, or even violate the Law of Environmental Impact Assessment. For example, the city EPB was left out in the selection and decision-making process for the selection of a site for establishing the 'Automatic Industry Economic and Technological Development Zone'. Normally, a project of such a scale should be reported to the local EPB for an EIA.[37] Yet this legal procedure was simply ignored.

Furthermore, Xiangfan EPBs do not have strong enforcement authorities. Xiangfan faces serious water pollution problems: the Gun River, a branch of the Han River, but also the Man River, Tangbai River and Xiaoqing River, suffer from serious pollution.[38] The lack of pollution-control measures by industry is the main cause of the pollution. Yet these enterprises were weakly enforced.[39] Knowing that each represents a non-negligible amount of pollution in the local economy, the polluting companies are unwilling to establish treatment technologies for discharging wastewater. In such a situation, the EPBs actively utilized public consensus to promote pollution control.[40] From this perspective, the EPB recognizes the GH as a partner.

The organizational link between the EPB and the GH was based mostly on Ms Yun's personal identity at the time when she worked as member of the XCPPCC. She was persistent in, and committed to, promoting the public's interests. The acceptance of Ms Yun's personality by heads of the local EPB promoted acceptance of the GH as a partner. So her position in this linkage proved critical. And the interaction between the two counterparts is best described as by personal ties.

Linkages with media

The GH works closely with the media, and some of its members are even journalists. In this way, the GH has favorable connections with the local media, such as radio or major newspapers, who joined the GH in supervising local environmental pollution.

The organizational links built between the GH and the media institutions have shown a dynamic change. At the beginning, personal ties and close informal interactions were needed when contacting individual reporters. Along with the development of the GH, it drew more and more attention from media professionals. As reported by staff of the GH, 'at the beginning, we had to invite journalists ourselves, but after more activities were conducted, more people started to know us. Nowadays, journalists come to interview us.' This change implies that building organizational links became easier after the GH had more social leverage.

At the same time, such links can also result from the local media's increased demands for reports. In the past several years, Xiangfan's news media have witnessed a growing interest in publishing environmental reports, a tendency also demonstrated on a national level.[41] Reports that are related to environmental protection, ecological improvements, and related topics are hot issues in the local media. Naturally, activities of the GH are of strong interest to them as well.[42] Gradually, building personal networks with media personnel began to be less important than before.

Linkages with ENGOs

With very limited local NGO partners in environmental reform, the GH actively seeks support for developing its activities and programs. Intimate organizational connections were built with 20 other fellow ENGOs, at both national and local levels, such as Friends of Nature of Beijing, and Nanjing's Friends of Green Environment.[43] These links have proved an important resource for the GH to develop environmental reforms locally. They not only exchange useful information, but also broaden GH's influence to a larger audience. The GH developed its activities by collaborating closely with the groups.

The organizational links between the GH and the ENGO community were based on formal relations between the two. Although Ms Yun acted as the main person linking the two counterparts, her role could have been played by other staff of the GH. Thus the nature of linkages was determined largely by the profile of the environmental group and not the actual persons involved. More specifically, building partnerships with other ENGOs is based on an equity principle, from which both counterparts benefited. For instance, the GH cooperated with Friends of Nature to launch programs in Xiangfan. On the one hand, such a cooperation program constitutes external support to the GH, which is a young educational NGO. At the same time, such arrangements were, for Friends of Nature, fulfillment of its own projects.[44] Staff of Friends of Nature went to Xiangfan three times to give training courses to teachers, when the GH was still in an embryonic stage. Friends of Nature also enabled GH staff to participate in their training courses outside Xiangfan.[45] In addition, the Institute of Environment and Development provided seminars on 'environmental teaching materials' in May 2004, to 33 schools and institutions, and one experimental group was chosen among others as an experimental target.[46] These organizational linkages added to the GH's experiences and efficiency in educational programs.

The GH also shares useful information with ENGOs. The GH is regarded as a respected member among other established ENGOs in China. It receives newsletters from other ENGOs and shares its own progress and information with them.[47] The GH's advocacy could be disseminated by posting protesting articles on ENGOs' websites, in newsletters and via publications, thus reaching a broader audience. For instance, articles written by GH members protesting against the policy of the South–North River Diversion were posted on the Friends of Nature website, and gained broader access to environmentalists in China.[48]

Furthermore, the GH has also been an active participant in common activities of the ENGO community in Beijing. In August 2005, the GH attended the 'Journalist Salon', which gathered together many environmental activists and was organized by the GEV and the Green SOS. At this occasion, the leader of the GH was provided a chance to give a speech introducing the activities carried out by her organization. Seeing Beijing as the central place where ENGOs, international donors, and environmental activists gather, the influence of the GH was greatly enhanced after contacting Beijing's ENGO community and delivering a speech at the 'Journalist Salon'.

Links with private enterprises

The GH also built a very limited number of links with private enterprises. As the social leverage of the GH grew, private enterprises became interested in cooperating with the environmental organization. The GH, however, did not show the same enthusiasm towards these enterprises as to its partners. It had no concrete plans to construct collaboration with private enterprises, and did not have complete faith in them. This was partly the result of the leaders' lack of knowledge of the operation of social organizations; they feared that their NGO would lose its 'not-for profit' nature through cooperation with private companies.[49] Besides this, Xiangfan's private economy was very weak, and considered by the public as not very reliable. So the GH leaders found it risky to construct organizational links with private enterprises, worrying that the GH's own public image would be endangered.[50]

In this case, only limited cooperation was developed between the GH and private enterprises. And the reason for collaboration with companies may not have been directly linked to environmental ideologies. Providing financial aid as a charitable deed was the prevalent form of cooperation. The GH received 5000 RMB (approximately US$750) from Xiaobai Xiang Company to print educational flyers, which contained an advertisement for the enterprise covering half the flyers.[51]

Inside the GH, the attitude towards market actors was not completely unanimous. With whom, and to what extent, to build cooperation were often issues of discussion. It can be estimated that with the development of the private economy and the change in leaders of the GH, collaboration with private enterprises will probably increase in the near future.

Campaigns for environmental reform

To demonstrate the roles of grassroots activists' efforts in promoting Xiangfan's environmental reform, I will present three cases of collective action. The first two cases focus on local reform, with different actors involved in each case; in the third case, a campaign led by the GH is presented, showing the tactics used to articulate protest to a neighboring provincial government.

Environmental educational program promotion

In recent years, local EPBs gradually became aware that the public is a partner in environmental protection. After the 'South–North River Diversion' policy that started in October 2003 was implemented, the local EPB decided to control domestic water pollution with stronger force.[52] The EPB is particularly determined to incorporate the public in its work to reduce domestic wastewater. Because the public's environmental consciousness is relatively low,[53] raising public awareness of the continuous deterioration of water quality and pollution becomes very important.

The GH successfully changed this situation by organizing coordinated actions with the Friends of Nature and two government agencies—the city EPB and the Bureau of Education. As a result of this effort, environmental education was institutionalized in the local educational system.

Environmental education conducted by local EPB

For a considerable time the EPB has been the only actor in environmental education. As it lacked materials and was not very systematic, the educational program initiated by the local EPB has not yet produced many positive results. With a weak capacity, the local authorities could hardly carry out effective environmental education to the masses.

Xiangfan environmental protection agency mainly relies on the government environmental educational center to conduct educational activities. The focuses of the Xiangfan EPB include mainly mass campaigning and publishing environmental reports.[54] In addition, teaching environmental law to political leaders to enhance their environmental consciousness also constitutes a main part of environmental education.[55] Mass campaigns were organized on special days, or during a certain period, such as 'Environmental Protection Day', 'Environment Protection Month', 'Festival of Environmental Protection', and so on. These were sometimes also demanded by MEP, which made them compulsory. For example, 'Environmental Protection at the Turn of the Century' is a campaign originally required by MEP. Each city-level EPB had to conduct a number of activities that have a certain social leverage. Yet such campaigns lasted only a short period of time. After they were officially finished, the consensus-building on environmental protection ceased. However, since such occasions are very limited and sporadic, the effects did not trigger a higher level of public environmental consciousness. Emphasizing environmental concerns and publicizing related knowledge through the media comprise another main activity by Xiangfan EPBs. The local newspaper and TV stations are the main media utilized. Posting advertisements on streets is another mechanism to disseminate environmental education. Through the publication of these articles and broadcasts, the EPBs expected public consensus to emerge concerning environmental protection.[56] Yet these activities did not take place continuously to mobilize public consensus.

In all, local EPBs had weak environmental educational capacities. They could not cover the 12 administrative regions, leaving space for ENGOs to fill the gap.

Construction of a campaign network

The GH acted as an important structural linkage, coordinating various actors. Under Ms Yun's efforts, cooperation was reached among three organizations: the EPB, the Bureau of Education, and the GH. On the one hand, through this link with government units, the GH acquired a small amount of support, such as transportation and small financial donations. On the other hand, environmental education was important to the two government agencies. By sponsoring this joint program, both government organizations could achieve political credit. In the process of initiating such cooperation, the role that Ms Yun played was critical. Having had connections with both EPB and the Head of the Bureau of Education, Ms Yun bridged the gap between EPB and the Bureau of Education with little difficulty.

In the implementation of public education, the World Bank covered most of the expenses. The GH, as the main actor in the program, completed the majority of the work: it invited professionals from the Friends of Nature, designed the education programs, organized participants, arranged venues, and so on.[57] For GH, the most essential resource needed for implementation of the educational program was authority, issued by the government. Under Ms Yun's design, in the process of organizing this program, the Bureau of Education and EPB jointly promulgated a document, 'Notification to Launch Environmental Education in Middle School, Primary School and Kindergartens'.[58] Through this document, the GH could mobilize participants smoothly. Talking about the document issued, Ms Yun remarked: 'it is a powerful sword for us ... no need to worry about getting participants.'[59] When conducting the training course, both political leaders of the two government agencies participated in important events such as the opening ceremony and 'graduation day'. They gave speeches, gave out certificates to participants, and thus brought authority to the program.

This program was a success. Five hundred teachers who came either from the city or rural areas were trained. Lectures were organized in 80 schools from early 2002 to April 2005.[60] A great number of students and young children were given environmental education, through which their environmental awareness was raised. The environmental program was also institutionalized in the local educational system. Schools are now required to install environmental courses and thus receive lectures provided by the GH. Later, this program was established as a citizens' educational program by the SCCO, as an experiment that was started by four residential communities and which should be broadened later on to include the whole city.[61] In this way, the GH could lecture widely on its knowledge of environmental protection, and also taught the public how to protect itself when its rights are affected by environmental pollution.

The success of the GH in initiating and implementing this educational program is based, to a significant extent, on GH's strategy to initiate coordinated actions with government authorities. In the context of Xiangfan, this seems very necessary for launching NGO's projects. By utilizing personal networks, GH acted as a broker, building structural links and institutionalizing environmental issues in policies.

Initiating an environmental governance model

The GH's role in local environmental reform can also be identified by the efforts it made in initiating a local ecological experiment in Wushan County of Xiangfan. The GH played an important role in promoting communication and coordination among a number of groups: actors from three different administrative levels of the government (city government, county government and villagers' committee), NGOs, and scholars. This was especially significant as efforts from various actors were required.

Because of their contacts with the GH, Beijing's NGO professionals became interested in launching their projects in Xiangfan. Among them was the Green Cross (GCR), an ENGO conducting environmental projects in rural areas. Through preliminary investigation, GCR discovered that the villages in Wushan County could very easily be developed in a sustainable way. The GCR decided to launch an experiment that would change Wushan County into an ecological village. This experiment consisted of a comprehensive system of projects aimed to reconstruct the village in an ecological way, for instance by establishing clean energy (marsh gas), improving the local environment, recycling rubbish, recycling telluric water, promoting ecological food and ecological tourism, and so on.

As the GCR feared difficulties in trust-building with the local government if they had not been introduced, the GCR did not intend to establish organizational links with the local government directly on its own.[62] Knowing that the GH had a preliminary plan to conduct an environmental education program in this county, the GCR joined GH, along with scholars from social sciences, to develop supportive ideas for the project. The expert team was composed of two men: Mr Gong, research staff of the Center of Environmental Development in the Chinese Institute Social Sciences, and Mr Tong, then a post-doctoral fellow in the NGO Research Center of Tsinghua University. After investigating the conditions of the villages, they were able to offer convincing arguments on the possibilities of launching such an ecological governance project.

Coordinated actions were based on the construction of networks among the participating actors. The GH was accepted by local villagers, and also acted as a representative for them, because it recently had carried out an educational program in this village. The SCCO, represented by Mr Zou, was a partner of the GH in this program. It sought to develop a new program in order to fulfill its task of promoting citizenship construction projects. By sponsoring and providing means of transportation, it planned to support the GH's educational program in Wushan County. The presence of the Vice-Director of SCCO, Mr Han, illustrated the importance of this project; in case of success such a project would serve as a model for a city scale. This indicated the supportive attitude of the city government to the project.

The Yanhe Villagers' Committee, represented by Mr Min (Party secretary of the Villagers' Committee), aimed at gaining better livelihood through both local ecological and economic activities. During Mr Min's efforts, Mr Yu (Party secretary of Wushan County Government) participated in the project as the top political leader and decision-maker, because the implementation of such reforms concerned collective actions of villagers, and financial support and policy subsidies from local officials at the county level. For both organizations, the success of this project would ensure them political credit.

The construction of an ecological village was a trust-building process that includes careful network-building. The GH acted as a broker, linking the Yanhe Village and the Wushan County governments that originally had no direct administrative link with SCCO, to the SCCO. The role of Ms Yun in this network formation is significant. Her background facilitated trust-building among the actors. Ms Yun, although a retired member of the XCPPCC, was trusted to still have the same commitment to serve the public as she used to in this political organization. Through her, the GCR was introduced and presented, and was received as a reliable partner in a short period, although the idea of NGOs was not very familiar to Wushan County. In this way, personal ties were constructed between the leader of the GCR and local officials.

Yet the organizational link appeared not to be very solid and lasting, and it was rather vulnerable due to the personal nature of the links. The GH's coordinating role did not last long in this project. After the cooperating network was built, the GH ceased its participation, for numerous reasons. However, the split was caused mainly by conflicts in informal occasions between Ms Yun and the leader of GCR. With the personal relationship disconnected, the GH was not present in the network any more. Fortunately, the loss of the GH did not affect the whole coalition, largely due to the construction of personal ties between Mr Sun and Mr Yu, based on the latter's charisma and reliable operations. Furthermore, the experts' arguments convinced the political leaders to begin systematic reforms. The local county government decided to support this project by giving subsidies for building marsh gas sinks.

Until the end of 2004, one year after the launch of cooperation, the two experimental villages greatly improved their sanitary conditions and increased the local ecological economy by 15 per cent on a per capita basis. Via the ecological experiment, the Yanhe Village began to show signs of sustainable development. The success of this experiment was promoted by the SCCO to the city government and it is under consideration to institutionalize this project in other counties in Xiangfan.[63]

Campaign protesting river pollution

As a conservative group under the restrictive government of Xiangfan, the GH seldom provokes local political authorities. In order to promote environmental reform programs, it usually is very careful when initiating its projects. Nevertheless, the GH is active in representing local environmental rights in advocacy activities towards political authorities in other provinces. Acting as a leader, the GH directed a campaign towards the Henan Government, which was primarily responsible for the existence of a polluted river running down into Xiangfan. The GH organized a movement network with the XCPPCC, the local media, environmental activists, and NGOs in Beijing. Because of this, a strong advocacy coalition was formed. Public opinion on this issue was also mobilized. A broad audience was reached, which even had an impact on the central government. The collective actions mobilized by the GH successfully exerted pressure on Henan Province, and facilitated measures taken to clean up the trans-boundary river.

Background

The Tangbai River is the longest branch of the Han River, and is located in the middle of the Han. The Han River diverts into the Tang River (in the eastern part) and Bai River (in the western part). Xiangfan is located in the lower branch of the Bai River, which runs from Henan Province to Hubei Province. The Bai River runs through Nanyang City of Henan before it enters Xiangfan. Its branch is 328.2 kilometers long, 300.2 kilometers located in Henan and 28 kilometers in Hubei. Zhaiwan Village is located on the bank of the Bai River and borders Henan Province.

The water quality recently appears to be worse than grade V.[64] The pollution is suspected to be caused by paper factories, which are located along the river banks in the Nanyang area (in Henan Province). As the heavily polluted water carries phenol—a chemical that may cause cancer in humans—the Bai River is reported to be the main cause of over 100 deaths of young and middle-aged villagers in Zhaiwan Village—a village along Tangbai River in Xiangfan (Map 5.1).[65]

To Xiangfan EPB, this issue of trans-boundary pollution made it impossible to deal with. Although the Xiangfan EPB constantly consulted with the Nanyang and Xiangfan environmental protection agencies, no strong measures were taken to diminish pollution caused by industries. Furthermore, during the process of consultation, conflicts sometimes occurred.[66] Fearing that this issue would lead to negative relations between the Henan and Hubei provincial governments, the Xiangfan city government did not report these problems to higher administrative levels. The serious pollution was barely controlled for a

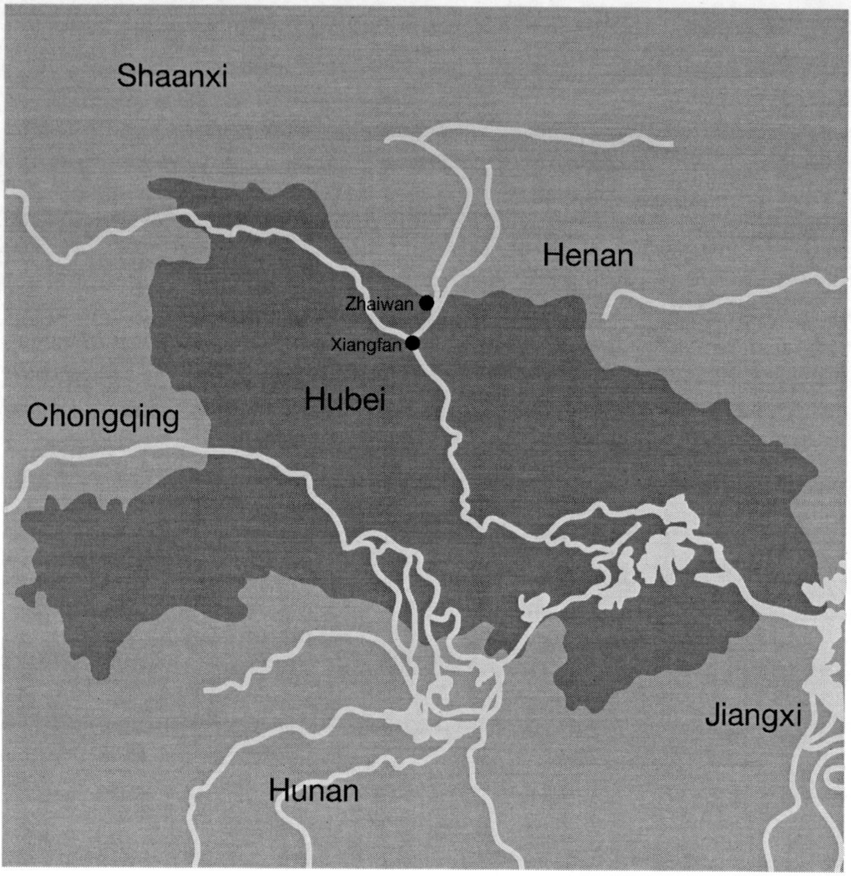

Map 5.1 The location of Xiangfan and Zhaiwan Village

long period of time. What is more, the Xiangfan authorities took a passive attitude, not allowing any government units to report on, or deal with, this issue. Because of this, the local EPB could express few opposing ideas on this issue to the public.[67] With the information on the Tangbai River water quality being confidential, news reports on this matter were banned as well.

Discourse construction

The GH found, through several investigations, that this situation started to worsen after 2004.[68] Through an informal discussion with the EPB, the GH acknowledged the sensitivity of this issue, and the weak capacities of the Xiangfan EPB in dealing with it. With this information, the GH carefully developed suitable strategies to stop the city government keeping the issue secret. Mobilizing collective consensus was adopted as the main strategy.

In April 2005, an 11-year-old primary school pupil, Duan Zekun, wrote a short essay entitled 'I Wish to Swim in the Tangbai River' after participating in the GH's educational program. In this essay of around 500 words, he expressed his wish to swim in a clear Bai River, and exposed the polluted state of the river. He pleaded with the Henan Government to cease the paper factories' pollution. With a keen knowledge of local politics, the GH encouraged the essay to be sent directly to Li Keqiang—the top official of the Henan provincial government—which started the construction of public consensus toward it.

The GH also collaborated with the Propaganda Department of the XCPPCC and collectively developed several news reports on this issue. As the XCPPCC had access to information on water-quality levels, it provided the GH with materials that could be disseminated to various media. The two institutions also created a video clip called 'The Village that Polluted River Runs Through', showing the situation of the pollution. This video was given to the GH's members who worked in news agencies. Soon, similar articles were published by several provincial newspapers of Hubei.[69] In this way the pollution of the Tangbai River was put on the public agenda on a provincial scale.[70] The GH did not stop the media campaign at a provincial level, and even contacted its own key members in the capital, who had the news released in national newspapers using their personal relationships. Two major national newspapers, the *People's Daily* and *Legal Daily*, reported this issue.[71]

By contacting NGO actors, a new tide of protest activities was organized by the GH. In May 2005, the GH invited volunteers from the Friends of Nature to investigate the pollution in the river. In January 2006, two well known environmental activists from Beijing, Ms Wang and Mr Ma, were invited to visit Zhaiwan Village, also called the 'cancer village'. The news reports soon spread to some influential media, including the main activists' websites.[72] Furthermore, these environmental activists also published their articles on the websites and in the media of related international ENGOs.[73]

Relying on mass media, the GH facilitated consensus formation on environmental protection, especially concerning water-quality protection. In one talk show program of the XCPPCC, the GH members talked in a lively way about their investigation into the Bai River, and reported the serious pollution.[74] With both local and national media attention, this issue was effectively disseminated to the public.

From public to political consensus

Through its media campaign, the GH built public consensus at the city level. This marked an improvement from the strict control of Xiangfan authority on this issue. By organizing activities, it widely mobilized public concern for water protection and indirectly disseminated information about the polluted TangBai River. Although no direct action was identified, the public concern for water issues grew.

Moreover, the GH inspired collective identity formation, which is less vulnerable to political constraints. Fearing that it might induce conflicts in local politics, the GH tried to avoid directly announcing the problem of Zhaiwan Village publicly. As an alternative, general values of environmental protection were expressed and elevated. The GH attached the protection of water to the value of 'family', a basic value cherished by Chinese people. The Han River is called the 'Mother River', so the relations between ordinary citizens and the protected river are strongly portrayed to be like those between a mother and her children. The GH adopted this metaphor from the Youth League, who first used it around the country as its slogan. The GH, however framed this metaphor according to a local context. It naturally describes the relationship between Xiangfan citizens and the river that runs across the city—children cannot be separated from their mother, and to protect one's mother is a child's obligation. Using understandable language, the GH amplified values of family and future generations to build consensus among the public. The framed values were strengthened by local culture, deeply rooted in Confucianism, which stresses the idea of family.

As a consequence, the collective actions organized by the GH directed MEP's attention to this issue. In September 2005, special officials were sent to investigate the Tangbai River in Henan and Hubei provinces.[75] Meanwhile, the polluting factories were closed and the Henan provincial officials decided to take control of pollutant sources along the Bai River more seriously. All in all, the campaign was a success. Mr Wang, Vice-President of Xiangfan EPB, remarked that the Henan provincial government had begun to take more measures on pollution control on the Tangbai River since Duan Zeshen's letter was sent to it.[76] The Xiangfan city government also started to provide financial support for the Zhaiwan villagers, and constructed a new well as a water source.

Conclusion and discussion

The case of the GH represents the force of an ENGO in a typical Chinese city that is undertaking an economic transition. Although the government's attitude toward civil society organizations was repressive, the GH managed to play an active role in improving environmental reforms by utilizing its personal and organizational networks. For the first time, GH represented the public in local environmental governance: environmental programs were institutionalized in policy-making. In the campaign on the trans-boundary river, the GH demonstrated that environmentalists could form a strong force to improve environmental policy, when political institutions did not provide channels to solve pollution control. However, being a conservative group in a repressive setting, the GH rarely organized adversary environmental campaigns on local issues in the twin city of Xiangfan. Only in the campaign that protested against the Henan Government did it act as a leader in articulating advocacy.

Looking at the networks that were built among movement actors, it can be recognized that government agencies from inside the political institutions were a primary force in promoting environmental transformation. Informal connections and networks of the GH were built and utilized to construct a coalition among movement actors, government actors, and other actors from outside the existing political institutions. The role of these personal networks could not be replaced by organizational networks. This demonstrates the prominence of personal networks in this social context. By using personal networks, solidarity and identity of environmentalism were formed and disseminated among the movement actors. Thus the GH could transcend boundaries among organizations.

The significant role that personal networks played in this context was rooted in the local socio-political structure. With the government dominating environmental governance, it is difficult to build organizational links without using informal mechanisms. Being a social organization that is not institutionalized in the political system, the GH had difficulties in accessing the hierarchical political system, and in gaining trust from its potential partners. Personal networks helped the group to articulate the environmental interests of the public and, to some degree, to participate in the environmental governance of Xiangfan. It should also be noted that the GH consciously complemented the government's responsibilities in its activities.[77] The ENGO adapted itself to the needs of the local environmental protection agencies and actively gained room for its development.

In this case, personal networks appear to be an effective mechanism utilized to mobilize environmental improvements. The ENGO, containing a social elite who had personal ties with officials, played a key role in the environmental reform processes. And personal involvement appeared to be a crucial strategy to facilitate communication and coordination among the movement actors. Thus it appears that the adoption of informal mechanisms can hardly be replaced by formal interactions of organizational networks in Xiangfan, and this continues to be the case.

Furthermore, the GH's organizational structure also shows that personal networks play a strong role internally. Personal connections are important not only in linking political institutional actors, but also in constructing and developing a civil society group. Although some formal elements were constructed in the GH's internal organization, the GH still relies on personal networks. As shaped by Xiangfan's past, the reliance on *guanxi* is rather strong. Intimate connections were utilized to exchange resources and power in organizing this ENGO. As a consequence, the utilization of individual ties will probably continue for some time under the influence of the local culture and current political system.

6 Growing environmental activism in Shanghai

Shanghai's socio-economic context

Political and economic context

Located at the mouth of the Yangtze River, Shanghai is the largest and most populous industrial city in China. It has a population of 17.8 million inhabitants and an area of 7823 kilometers. The city has experienced rapid economic growth since the reform era in 1979, and has developed into a financial, commercial, and international business center in China. Its GDP per capita was around US$4500 in 2005, which ranks Shanghai 12th among all Chinese cities.[1]

As the second largest city in China, Shanghai has been the largest contributor of tax revenue to the central government compared with other Chinese provinces and municipalities. Before the 1990s, 83.9 per cent of its tax revenue was submitted to the central government (Shanghai Chorography Editing Office, 1997).[2] This came at the cost of severely crippling Shanghai's infrastructure and capital development. The municipality's importance to China's fiscal wellbeing interfered with China's economic liberalization, which started in the southern provinces such as Guangdong during the mid-1980s. Only after 1992 did the central government under Jiang Zemin, a former Mayor of Shanghai, begin to reduce the taxes Shanghai had to pay to the central government. This encouraged both foreign and domestic investments, and promoted the city as the economic hub of East Asia and the gateway to the Chinese interior. Since then, Shanghai has experienced continuous economic growth of between 9 and 15 per cent annually, leading China's overall development.

The private economy has developed rapidly in Shanghai over the past decade,[3] and reached 42.5 per cent of the local economy in 2005. It has become the main sector that creates working opportunities in Shanghai: in 2005, 58 per cent of the labor force in the city was in the private sector (Shanghai Municipal Statistics Bureau, 2006). With increasing investments, the scale of enterprises also grows quickly. The average investment of enterprises is around 710,000 RMB, which is higher than in the neighboring Shandong, Zhejiang, and Jiangsu provinces (Zhang, C., 2001).

Accumulating wealth in Shanghai was important not only for fulfilling the responsibilities of government agencies, but also to individual leaders. These political positions have traditionally been a stepping stone to higher positions within the PRC central government. Four Shanghai mayors eventually went on to take prominent central government positions, including former President Jiang Zemin and former Premier Zhu Rongji. This contributes partly to the restrictive attitude of the political authorities.

The municipal government has given special attention to the regulation of social organizations. In the bureaucratic structure, the Shanghai Bureau of Civil Organizations (SBCO) was established out of the Department of Social Organization Regulation in 1999, with the task of controlling social organizations.[4] Since its establishment, the SBCO has been sufficiently equipped with facilities and financial input. 10 million RMB was invested in the construction of this agency, which places it among the richest agencies across the country. Unlike similar agencies, which usually rank as departments in administrative ranking and are subordinate to the local Bureau of Civil Affairs, the SBCO is established as a bureau in the same rank, and thus has been provided with strong power. A special team, the Social Organization Investigation Team, which consists of 30 members, was established in the same year, and strengthens the force of this bureau in keeping control of dissidents. In addition to this well established agency, a special system of social security called the 'four level security mechanism' was developed to safeguard social stability, under the joint collaboration of the Bureau of Civil Affairs and the SBCO. This mechanism is composed of major government agencies in four administrative levels, and works in a comprehensive way so as to be alerted when social dissidents emerge, even on a small scale (Shen, 2002).

NGOs in Shanghai lack autonomy and independence from intervention of the local government (Yang, X., 2004). The government of Shanghai is very cautious in legalizing and allowing citizens' organizations, and hinders citizens in realizing the right to form associations. At present, political authorities prevent the development of NGOs by not providing them legal status. Civic organizations or interest groups in the legal form of social organizations are very difficult to legalize. When registering this type of NGO, professional or industry associations are preferred. In 2005, around 42 per cent of NGOs were professional or industry associations.[5] Aside from this type, many NGOs are registered as non-profit enterprises (*minfei*) to provide services in education, public health, and so on.[6]

The government is also limiting free speech. Local media are conservative in reporting news that may not be politically innocent. As the government strictly controls news reports, the media enjoy only limited freedom in reporting issues. The press and media houses normally know their limits, and appear conservative in reporting on topics and issues that may cause emerging civil unrest.[7] Besides this, the Shanghai municipal government also strictly supervises the content of the internet. They limit the utilization of cyber cafés in the city, in order to prevent people from having access to subversive online content.[8] Internet censorship is also seen in universities. Under the direction of university authorities, student monitors are selected who secretly tame the web.[9]

Local culture

Shanghai is a city where the use of *guanxi* and personal connections is less prevalent than in other Chinese cities. This is closely related to the fact that it recently became the biggest financial city in the Far East. The city has grown dramatically since the nineteenth century, because its strategic position at the mouth of the Yangze River made it an ideal location for trade with the West. It became an international port and the financial and commercial center of China in the 1930s. It also developed into one of the most important industrial bases in China before the Second World War. After the Open Door policy began in the early 1980s, Shanghai regained its position as financial and commercial center, with the return of foreign banks in the new Pudong Financial District, and the establishment of a modern stock exchange market. Thus Shanghai became the most international city in China, and was greatly influenced by Western culture (Chen, 2001; Wang & Tong, 2005). Its comprehensive infrastructure and efficient legal environment attract a considerable number of multinational corporations, which build their regional headquarters, or research and development centers for Mainland China, in Shanghai. In 2000, half of the world's 500 top companies had investments in the city.

In this international city, rules, contracts, and regulations that are adopted from business culture have become deeply rooted in the local socio-economic environment. In individual interactions, economic profits are more readily pursued than the building of personal connections. This differs from fraternal loyalty—a sense directed by personal connections particularly recognized in Beijing (Li, C., 1996). According to a survey on managers working in state-owned enterprises in Shanghai, a strong tendency to individualism was found. Located on China's east coast, these features pose a major difference from the metropolises of Chongqing and Lanzhou, located in southwest and mid-western China, respectively. Some scholars argue that this shows that Shanghai citizens have achieved little heritage from Confucianist values (Ralson *et al.*, 1996). They conclude that personal networks have little influence on the social organization of inhabitants of Shanghai.

All in all, traditional Chinese culture has relatively limited impact on the behavior of individual citizens in Shanghai, and the westernized cultural context relies less on personal ties than on informal mechanisms for interaction.

New communication means in the information age

As Shanghai is a modern city that is developing rapidly, the internet is widely used.[10] Shanghai citizens are active users of the internet, and frequently access information through the net as an alternative to newspapers and TV.[11] According to a survey conducted in 2005, 4,410,000 Shanghai inhabitants use the internet, which accounts for 26 per cent of the whole population.[12] This is about 50 per cent more than the current average percentage of urban inhabitants who use internet across the country.[13]

Communication software tools provided by the internet, such as email, electronic forums, and other forms of interaction software, are widely utilized. With the net, greater potential for communications is available in a cheap and effective way. Cyberspace helps Shanghai citizens to advocate their ideas and communicate with like-minded people. Moreover, the internet proves to be an effective method in terms of increased speed, reduced costs, and persistent accuracy, all of which are important for communication among activists in the city of Shanghai, where face-to-face interaction is costly in terms of time and money.[14] In such circumstances, using group emails and posting information online have become the preferred methods of communication.

With the prevalent usage of the web, various social groups that use the net as their dominant method of communication have recently emerged. According to a report on the administration of NGOs in Shanghai, SBCO is very concerned about the associations that communicate and operate through the net, which is growing rapidly in Shanghai (Shen, 2002). These groups have a broad constituency, are often weakly organized, but can still attain a certain degree of social leverage.[15] The political authorities keep a close eye on the development of this kind of social organization.

Environmental governance

Shanghai has recently received criticism for its environmental governance. Historically, serious pollution resulted from the economic boom in the late 1970s. The municipality's focus on improving the environment started after it gained more financial independence in 1990. Since then, billions of yuan have been spent cleaning the air and improving water quality, with great emphasis on the establishment of physical infrastructures.[16] A specific policy committee was established in order to implement and enforce local environmental policies and regulations. In a decade's time, Shanghai greatly improved its environment: SO_2 emissions were reduced by over 50 per cent, and more than 80 per cent of the enterprises reached the national pollution control standard.[17] At present, water quality, air pollution control, solid waste disposal, and city greenery are the top areas of improvement. The municipality has high hopes of becoming an ecological city by the year 2020.[18]

Shanghai's environmental governance is characterized by large enforcement capacity, effective collaboration with other non-governmental actors, and low involvement of the public in clearing the environment.

Environmental enforcement in Shanghai

The Shanghai Environmental Protection Bureau (SEPB) and its subordinate environmental agencies are strong in environmental protection, as demonstrated by their significant administrative force.

As the top environmental protection agency, the SEPB is well facilitated, and is capable of exercising authority. More than 2000 staff members are employed by the SEPB, including district inspectors and enforcement personnel, which ranks the city higher than one would find in the much larger and more populous provinces in inland China. Furthermore, the SEPB has been able to collaborate with other government units of Shanghai in a flexible fashion. The district police station, the urban planning bureau, and the security bureau are mobilized by the SEPB to strengthen its enforcement. What is more, special offices are often established for specific projects.[19]

In order to address environmental issues, the municipal government established a concrete plan. A new round of the 'Three-Year Plan of Environmental Protection' was launched in 2006, after the first two were completed in 2005. To ensure the implementation of this plan, a specific mechanism was established in their environmental policy-making so as to coordinate the interests of different government units and facilitate communication among them. In order to implement its ecological action plan, an 'Environmental Protection and Ecological Construction Coordinating Committee' was established in 2003, which proved to be a unique governance innovation across China.[20] Composed of 30 municipal agencies and all the district/county governments of Shanghai,[21] this committee marks the first city environment committee across the nation to be led directly by the mayor and three vice-mayors. The committee is especially efficient in policy implementation, because authorities in both horizontal and vertical levels are mobilized. It suits the structure of Chinese bureaucracy,[22] through which conflicts may rise between *tiao* (vertical) and *kuai* (horizontal) units. What is more, this mechanism also boosts the capacities of environmental protection agencies, which may have been weak in implementing policies alone. Seven specific working groups that govern six environmental issues were established, including water management, air pollution control, solid waste control, industrial pollution control, agricultural pollution control, and urban greenery. The SEPB and its subordinate units play a role in monitoring projects, coordinating actors and conducting evaluations. In this way, different government agencies have clear responsibilities for implementing environmental policies.

The environmental protection plan is characterized by using administrative measures, and the Shanghai Government favors administration by law (Office of Implementing Agenda 21, 1999). Under the MEP direction of strict adherence to law in environmental enforcement, Shanghai has developed a series of legal documents in recent years. Since 2000, more than 14 legal documents have been promulgated in five years.[23] In 2005, merely on the issue of suburban wastewater, five laws were promulgated, including Shanghai's Regulations for Promulgating Environmental Impact Assessment and Regulations on Market's Dirt Control.[24]

The SEPB has considerable regulating capacities, which can be illustrated by the large number of penalties it has given. In 2003, 454 factories were charged for their waste emission, and 566 were fined for not complying with project construction regulations. As for solid waste management, in that year 2230 garbage-sorting stations were built, and another 250 stations for exchanging waste and used materials. Ten temporary garbage spots along the Suzhou River and 764 temporary spots at the county and village level were pulled away.[25] As for the enforcement of environmental impact assessment (EIA) regulations, a high rate was reached by utilizing the principal of the 'three synchronizations'.

The SEPB is also equipped with considerable revenue in environmental enforcement. Since 1999, the Shanghai municipal government has been trying to invest at least 3 per cent of its revenue in environmental protection (Table 6.1). In 2004 and 2005, over 500 billion yuan was spent to realize its second 'Three-Year Action Plan.'[26] Besides this, Shanghai is also drawing interest from many international organizations as a focal point for experimentation on improving environmental governance in China. Technological research and training for civil servants have been initiated in the city, supported by international organizations such as the United Nations Development Programme, the United Nations Environment Programme, and the World Bank. Thus financial support worth billions of dollars has been donated to Shanghai.[27]

Shanghai also sets a higher norm than the average national level, with stricter environmental standards.[28] Shanghai's environmental standards—notably those governing water quality—are in fact higher than national standards. The Shanghai Municipal Health Bureau, for example, is exploring the idea of upgrading mandatory water quality standards to match World Health Organization and EU levels (although this idea has encountered some opposition from interests that find that hurdle too high to jump in the short term).

Collaboration with private actors

The Shanghai Government is active in promoting private sector participation in environmental governance. The business sector, made up of both private and state-owned enterprises, is engaged in tackling environmental challenges. Also, technical and financial support from overseas has been used effectively.

Table 6.1 Input in environmental protection

Year	*GDP (billion RMB)*	*Increase in GDP (%)*	*Spending on environmental protection (billion RMB)*	*Percentage of GDP spent on environmental protection*
1999	4034.96	10.2	111.60	2.77
2000	4551.15	10.8	141.91	3.12
2001	4950.84	10.2	152.90	3.09
2002	5408.76	10.9	162.40	3.00
2003	6250.81	11.8	191.53	3.06

Source: website of Shanghai Environmental Protection Bureau.

With the city's rapid economic growth, the municipality became a center for the most advanced environmental thinking and for cleaner production, creating a booming industry in environmental technologies (Economy, 2004). Take, for example, the water management market, where since 1980 international companies began to enter the Shanghai market. After 2000, local companies also became active in this market (Lee, S., 2006). According to the SEPB, there are over 1000 companies and 107 institutes and research centers engaged in protecting the environment, which brought in a gross profit of US $1.5 billion in 2004.[29] Since 2003, the SEPB began to form shared service platforms, including technical support, investment, marketing, information, consultation, and cooperatives with universities, relevant research institutes, and other service providers.[30] This agency has built working relationships with various actors to work on environmental protection. The municipal government's dominating role is also complemented by the work of government-operated NGOs (GONGOs), through which a number of private actors build working relations.

All in all, Shanghai is turning itself into a city suitable to live in, with blue skies, clean water, green land, and a sound environment. As Zhang concludes, Shanghai's economic development has facilitated the city to reach a developmental stage in which life-supporting ecosystems, in particular fragile ecosystems, or irreplaceable natural resources are kept intact (Zhang, M., 2002). Furthermore, according to one of the latest reports of the Chinese Academy of Science, Shanghai ranked first as a sustainable city among cities in 31 Chinese provinces (Research Team on Sustainable Development, 2005).

Public involvement in environmental protection

In comparison with the market actors, the role of the public and environmental NGOs (ENGOs) in Shanghai's environmental reform is small. In EIA processes, for instance, limited channels exist for non-organized public participation. As regulations and decisions are conducted by the SEPB without any public supervision, public interests have been constantly neglected in favor of bureaucratic and business ones (Lo *et al.*, 2000).

Despite the exclusive policy of the government, citizens of Shanghai do not remain passive in the area of environmental governance. With growing environmental awareness, the public has expressed enthusiasm in participating in environmental policies. For example, the official website for environmental protection was viewed 4 million times in only one year after the website was established in 2003. Also, 25,000 calls to hotlines were made in 2005 to complain about affairs related to environmental protection. However, in comparison with the active role that market actors and mechanisms play in environmental governance, the public and civil society are relatively poorly involved. Furthermore, through policy plans and projects, the municipal government encourages market involvement more than public. With few channels available, environmental volunteers are relatively passive in environmental protection.

Environmental movement in Shanghai

General introduction to ENGOs

In general, the emergence of environmental activism in Shanghai appears to have developed on a different track from that in Beijing, where ENGOs have a certain social leverage (Wu, 2002b; Lee, S., 2003). Shanghai's NGOs have begun to launch environmental projects only since 2000, when it became fashionable to contribute to activities to improve the environment.[31] The growing environmental activism in this cosmopolitan city is strongly deterred by restrictions applied to environmental organizations, mostly related to the registration of such groups.[32] Illegal groups can hardly survive in the local political environment.[33] A small number of ENGOs exist in Shanghai, solely because they are not confrontational to the government. Similarly to Beijing's groups, Shanghai's ENGOs lack professional knowledge (Lee, S., 2003).

In Shanghai, three types of group can be identified: GONGOs, international NGOs (INGOs), and grassroots voluntary organizations.

Shanghai's GONGOs are active in complementing government agencies within their responsibilities. Several play a strong role in Shanghai's environmental protection, which can be divided into three kinds.

- Traditional mass organizations, such as the Youth League, act as bonds to link grassroots NGOs with environmental agencies, and channel the public to participate in this process.[34]
- Organizations that are affiliated with environmental protection agencies, such as Shanghai Environmental Educational Committee, which conducts environmental education and is also in charge of mass campaigns and environmental educational projects of SEPB.
- Professional organizations that promote environmental protection in specific fields of activity, such as Shanghai Wild Animal Protection Association (SWAPA), which acts as a professional association dedicated to the protection of animals.

GONGOs initiate and implement their activities in a way that is strongly dependent on the administrative structure. Owing to the effective bureaucratic system of the municipal government, Shanghai GONGOs are capable of playing a prominent role among the various environmental organizations in the city.

The second type of environmental group in Shanghai is the INGO. These work mostly on issues of nature conservation. With a largely undeveloped green civil society in Shanghai, very few international ENGOs exist, and even fewer conduct projects in the city. The majority of international groups have their projects outside Shanghai and use the city as a base. An example of these groups is the Wildlife Conservation Society (WCS).[35] Shanghai's

INGOs work closely with international companies, particularly in providing environmental education to the public.[36] Unlike Beijing's INGOs, they are less involved in mobilizing local environmentalism. Recently, the Shanghai Oasis, a local grassroots group, was established with support of the WCS as part of its growing scale of activities. However, such efforts are limited. Also, the success of this facilitation can be attributed to the fact that the work of Shanghai Oasis is politically 'innocent'.

Currently, only a limited number of grassroots groups are specifically dedicated to environmental protection activities. Although their focuses are different, a general sense of environmental activism is gradually formed among grassroots NGOs. Grassroots NGOs committed to environmental protection are mostly registered organizations, which can be distinguished into three kinds: social associations, non-profit enterprises, and student environmental groups. At the same time, a small number of them are not registered and some are internet-based.

In general, Shanghai's grassroots NGOs have difficulties in getting legalized status. Because of the restrictive regulation of the municipal government, very few membership-based NGOs are allowed to exist. So they are often registered as non-profit enterprises. Very few social organizations specifically devote themselves to environmental protection: most of the ENGOs cover a variety of goals including green activism, such as religious associations and NGOs that provide social services. Currently, the number of ENGOs—primarily composed of social associations and non-profit enterprises—account for a very small proportion of the total number of NGOs in Shanghai.[37]

Aside from these types of NGO, student environmental protection associations are another type of grassroots group that were established in universities. The emergence of these groups was a result of growing environmental concerns among students. The groups function as a means for young people to articulate their common interests toward nature protection and environmental education. Currently, several dozen student environmental protection associations exist. The Shanghai Green Student Forum (SGSF) was formed especially to build coalitions between them.

In addition to registered ENGOs, web-based environmental groups are an important type of unregistered NGO in Shanghai. This type of ENGO is organized among individual citizens who share common interests, such as traveling, protecting animals, and conducting social charity activities. Green activism is a growing interest among individual citizens in Shanghai. In recent years, the cosmopolitan city has experienced a boom in various kinds of online group, through intense use of the internet. These groups consist of individuals who share interests in, among others, cars, pets, and traveling. The number of participants in this kind of web-based NGO is large.[38] They are characterized by a weak organization and a strong reliance on computer-based communication. This type of group is formed informally by individual volunteers. Because of the difficulty of finding a supervising institution, very few are registered. Most of these groups thus remain loosely organized.

Methodology

As grassroots NGOs are the main focus of this study, this type of ENGO has been chosen for this chapter. Three NGOs have been selected because they represent successful organizations in the local environmental community, and have a significant number of members and a longer history: the Grassroots Community (GC), WWF Shanghai, and the Shanghai Green Student Forum.[39]

In addition, the three groups show different types of legalized NGO. The GC is a registered social organization, WWF Shanghai is a web-based ENGO, and the SGSF is a secondary social organization that is registered within each university. The different forms of legal status indicate that the degree of political recognition differs among the groups, which may affect the degree of reliance on personal networks and lead to different tactics and strategies these organizations adopt within the environmental movement.

Grassroots Community

The GC is a legalized social organization established in 2000. The GC is embedded in the Zhijiang Xilu Sub-District, one of eight sub-districts of the Zhabei District. A population of 86,399 live in this area, which governs 22 Residents' Committees. The GC has built cooperative working relations with these Residents' Committees, which are the lowest level of administrative management and which implement polices made at higher levels. The construction of this relationship is made possible by the administrative reforms that happened since the mid-1990s, when the municipal government began to decentralize its authority to community-level district governments. These reforms enabled Residents' Committees to transform from just government representatives to self-governed organizations. In this process, democratic elections are initiated, which permit participation of citizens in policy-making. As a result, these residents' committees begin to enjoy a greater extent of autonomy, and are strengthened in implementing government tasks and provision of public welfare in communities.

In terms of annual budget and project scale, The GC appears to be a big grassroots NGO in Shanghai. Supported by a team of full-time staff, it has 94 members and a number of volunteers. Since its inception, several international foundations have recognized the GC's solid work, such as the Canadian Center for Civil Society Development and the Hong Kong Community Partners. Thus it has sufficient funding to implement its activities, including short (two- to three-year) and medium (five-year) projects.

The GC provides services to the inhabitants of Zhijiang Xilu District. With its practical working style, it has been successful in promoting welfare through various kinds of service, ranging from legal aid to job training for a group of village women who followed their working husbands by immigrating to Shanghai. Through concrete work at the grassroots level, the GC has won trust among residents. This provides the basis for launching environmental projects.

Since environmental issues are of more and more importance in the lives of citizens, improving their environmental awareness and rights in such issues has become an unavoidable part of civic training. In 2001, the GC developed its environmental project. Its general goal is to build a 'green community', in which residents are encouraged to take autonomous decisions regarding the environment in their community. This goal complements the GC's original goal of fostering citizenship and promoting the development of civil society. In the environmental projects that have been developed, environmental education is the main content of the activities at present. Residents are trained to be able to govern themselves in protecting their neighborhood's environment. Garbage sorting and recycling are set up in the community. Guided and directed by the GC, 'Neighborhood Environmental Protection Festivals', tours and seminars have been organized to raise the awareness of residents. With careful planning and considerable work by the GC, the Lingguang Garden Community's environment was improved and won a city level award as a 'green community' in 2003.[40]

Organizational development

The founders and leaders of the GC are three young persons, who were close friends in the Fudan University of Shanghai. Their charisma, knowledge, and insight mobilized a great number of followers, and promoted the development and growth of GC. Due to the fact that a large number of volunteers have a degree in law, members of GC tend to rely heavily on institutions in its operation.[41] In this way, GC was founded as a formal organization, which differs greatly from the majority of grassroots ENGOs across the country.

The GC established a larger number of rules and regulations for the management of members and for decision-making mechanisms. It has institutionalized a democratic system of rules regarding internal management (Figure 6.1).[42] Rules prescribing procedures and behavior in work are set out, such as regulations regarding arranging duties, registration, and overtime work. Particularly, the membership system of GC is well established. Members are endowed with rights regarding involvement in policy-making and management processes of the organization. The general members' meeting, which is held once a year, is the highest mechanism for determining the goals, strategy,

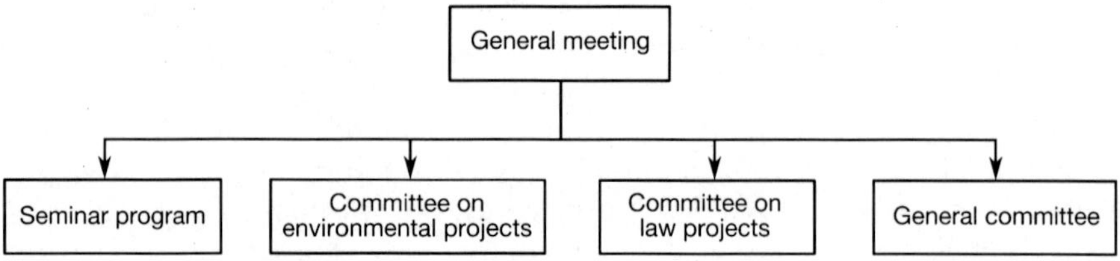

Figure 6.1 Internal management structure of GC

and regulations of the GC. In daily operations, a general committee made up of four or five key figures of the organization is responsible for implementing the decisions of the general members' meeting and for making decisions for GC. Project committees are also formed for every project, which enjoy high levels of autonomy. Each committee consists of project officers and core members, who collectively make decisions regarding the specific project. In general, individual participants in GC are strongly motivated to get involved in the organization's development.

The internal institutionalization of GC is effective in directing its operation and management. As a result, members of GC are well regulated and a large number of them are actively participating. For this reason, GC's internal management is appreciated and referred to by its fellow grassroots groups, and is taken as an example by some of them, such as Green Power and Shanghai Wild Birds' Association.[43]

Group identity

Looking at the composition of GC members, we see that they firmly rely on established institutions and rules in governing themselves. They have a relatively clear collective identity.

As passionate, young to middle-age professionals, the followers of GC are energetic, open-minded, and quick to adapt to new issues and problems. Living in the international mega-city of Shanghai, they are greatly influenced by globalization process, and are stimulated by the value of individualism. They are therefore less influenced by the culture of utilizing personal networks, and are less inclined to adopt *guanxi* as a strategy in realizing their goals.

To this group of individuals, participating actively in volunteer work is one means of fulfilling their ideals. Each of them has a strong belief in promoting civic development through their own efforts. Improving environmental protection constitutes one of the goals in their activism toward reducing social inequalities. Another aspect that individuals share is their need to meet like-minded people and make friends. In a city where inhabitants originally come from different places, meeting people comes naturally to the immigrants. More than 50 per cent of the activists are not local Shanghai people. Their desire to make friends and expand their social circle is satisfied by joining GC. In this organization, women volunteers account for a slightly larger proportion of the members than men. Members matching among themselves and becoming couples is not rare.[44]

Organizational links

Externally, GC has built rich relationships with various actors, including the NGO community, international organizations, and government agencies. Through their organizational networks, GC possesses social leverage and impact, which are less prevalent in other local grassroots groups.

The way that GC establishes its organizational links results from two aspects: its own internal organizational structure, and the way in which its counterparts operate. First, the institutions and rules in the GC determine that members and staff have their own responsibilities and obligations. Therefore specific staff are given the responsibility of contacting and sustaining professional linkages with other organizations. When building links between GC and its counterparts, the members and staff of GC are aware that they should not confuse personal interests with the organization they represent. Second, the organizations with which GC interacts rarely utilize personal networks, which also contributes to the formation of organizational links between them. Thus exchanging mutual organizational benefits becomes an important principle for the development and maintenance of organizational links.

Networks among NGOs

The GC has built organizational networks with local NGOs, as a strategy to acquire knowledge and strengthen its capacity. It has built rich links not only with local NGOs in the Shanghai region, but also with influential groups across the country.

The GC has learned a lot from experts from Beijing ENGOs, particularly how to conduct environmental educational projects. By visiting ENGOs and attending conferences held in Beijing, members of GC have built contacts with Beijing ENGOs, such as Friends of Nature, Global Village of Beijing, and Green Cross. Furthermore, this organization also worked closely with local influential NGOs. Through collaboration with the Shanghai Young Men's Christian Association (YMCA), an influential local NGO, GC built a strong volunteer team, particularly in environmental projects. The YMCA has a lot of experience with managing and training, and has provided training to volunteers of GC. Since most of their volunteers were professionals in fields not related to the environment, training provided volunteers with up-to-date and accurate environmental information.[45]

Furthermore, GC has also constructed links with international NGOs, with which it cooperates and from which it acquires professional knowledge. Cooperation with international actors was not only limited to financial issues, but also involved collectively organizing educational programs. Knowledge and information on specific programs are the major resources that the group gained. As this organization did not possess a team of experts in environmental projects, its program officers actively sought cooperation with local professional INGOs, which focus on specific issues.[46] In its educational program on the protection of wild animals, GC cooperated with the WCS, from which it learned how to design and produce exhibition boards.[47] For its program in genetically modified food, GC cooperated with and learned from Greenpeace.[48] This collaboration helps to compensate for GC's lack of professional and scientific expertise.

Through mediation with INGOs, GC was also facilitated to building links with other NGOs. GC was introduced to the Beijing New Citizen Education

Centre's (BNCEC) citizenship training program.[49] The two groups designed a 15-day training course for the GC green community program. Staff from the BNCEC trained the residential community and active community members. Through such cooperation, the GC program was strengthened.

Organizational links with government

Aside from an organizational network among the NGO community, GC has built organizational links with local government. It cooperated with administrative agencies at two levels—residents' committees and Zhabei District Government Environmental Protection Agency (ZBEPB).

In its early stages, GC focused mainly on one area of activity—providing legal assistance to inhabitants. When developing its activities, GC organized its activities closely with local communities—Zhijiang Xilu Community, the No. 710 Community.[50] Under the decentralization process of the Shanghai municipal government, the Residents' Committee of No. 710 Community was endowed with stronger autonomy than before, and was experiencing a transition, which was greatly supported by its inhabitants. GC's activities facilitated the participation of inhabitants in local affairs, and realized citizens' self-governance at the community level.[51] Underlying environmental concerns were expressed, focusing on the need to clean up the dirty, disordered living conditions of the community.[52] With the help of volunteers who were mobilized by GC, inhabitants realized their concerns and actively participated in the Residents' Committee's work. As a result, GC's work was welcome by the Residents' Committee, and the two actors began to complement each other in improving the community's environment.

The achievements of GC were soon recognized by district government agencies—the ZBEPB, which actively initiated cooperation with GC. Since 2004, GC was incorporated in ZBEPB's public education program. In contrast to the past, when such activities were launched purely directed by government agencies, GC and representatives of the local community participated in the organization of a mass campaign on 'Earth Day'. Recently, ZBEPB also tried to incorporate GC in its pollution enforcement activities. The social group helped to cope with small restaurants that produce serious solid waste in the local community, because it was believed to be influential in this area. Seeing that charging fees and utilizing administrative measurements cannot effectively stop pollution, ZBEPB began—through GC—to incorporate the public in its enforcement.

The development of cooperation between GC and ZBEPB brings opportunities for the public to be involved in the implementation and even, to some extent, the making of local environmental policies. In addition, through collaborating with the lowest administrative branch of Shanghai municipal government, GC could develop its projects with high efficiency. In this way, GC gained recognition from political authority, and thus to some extent was protected by this agency. Its freedom in initiating and implementing activities was considerably broadened.

Role of GC in local network

Under Shanghai's repressive political environment, very few coordinated efforts are recognized between environmental groups. Yet GC is located in the center of the organizational network of local grassroots ENGOs, because of the various types of resources it mobilizes through both personal and organizational links. It functions as coordinator of local NGOs by providing a platform for small groups to develop, and fostering growth of civic virtues among environmentalists. In general, GC plays an important coordinating role among grassroots environmental organizations in Shanghai.

The GC utilizes common members' affiliations to link other grassroots groups. The impact of GC attracts members from local groups, and through these common participants, it has been able to build organizational links with other ENGOs, such as WWF Shanghai and student groups. These organizational relationships, built on connections between individual activists, make it possible for GC to provide resources to others.

In general, through its organizational links with government, GC acquires a number of resources, including information and political protection. Further, it utilizes the organizational network with its ENGO partners for communication and coordination. In 2003, GC was given a position to participate in one official activity of the Youth League in Zhabei District, its supervising institution. Through the affiliations among common members of GC and other NGOs, GC released this news to other NGOs. In this way, members of WWF Shanghai gained the chance to be channeled to the government's environmentalism.

What is more, the fact that GC is concerned about multiple issues makes it easier to attract common members, and broadens its organizational links. GC's seminar program fosters ideas of civil society, with various topics that attract grassroots NGOs. A number of environmentalists have become members of GC after joining this program.[53] At the same time, the seminar also facilitates information to be released. In May 2004, when the campaign on the Nu River was very heatedly advocated by Beijing's NGOs (see Chapter 4), one of its activists came to Shanghai to build a wider network. The GC helped her to give lectures both at its seminar and at a gathering at another local NGO—Green Power. Dozens of environmental activists joined the seminars. Information about dam-building opposition was disseminated. After the lectures, the activist also successfully sold postcards and pictures, part of a donation to this campaign. In this way, Shanghai's civil society could be incorporated in national environmental campaigns.

Through GC, a fluid network has been built up among Shanghai's NGO community that shares a common interest in environmental protection. For security reasons, the NGOs and activists are very cautious not to build large-scale actions,[54] yet the efforts of GC in building and expanding this network are undeniable. What is also seen from the functioning of this group is that formal links between GC and other organizations, including governmental agencies, are of more importance than links between individuals. Personal connections are instrumental in building structural links between individual organizations. This kind of link is prominent, as some NGOs are far from fully developed.

Shanghai Green Student Forum

Students' green interest groups are prevalent in a number of universities in Shanghai. For young university students, experiences of 'voluntary work' are an important credit in job recruitment for the local market, especially within international enterprises. More and more students are ready to participate as volunteers in environmental activities. Among the 50 local universities, more than half have established students' interest groups dedicated to environmental protection.

Under the influence of the students' environmental protection associations in other cities, such as Beijing, Xi'an, and Dalian, the growing enthusiasm of students has resulted in the establishment of SGSF. It is an organization that promotes cooperation among student groups, through which student environmental activists can achieve greater influence among the public. Both SGSF and its member groups help to cultivate a growing pool of environmental activists, who might become future staff in ENGOs.

Organizational development

In their area of activities, SGSF and the small groups of students primarily conduct environmental education. The educational activities focus on, for instance, paper and battery recycling, both on and off campus. Some of the Students' Group could even integrate professional knowledge in their activities, and successfully promoted research in environmental protection.[55]

SGSF's internal organization can be examined at two levels: that of each small group, and the higher level of SGSF. At both levels, the organizational development shows relatively high levels of institutionalization.

Influenced by student groups in other cities, Shanghai's young students paid close attention to their internal organization. Since they were established, student groups built systems of regulations to direct their work. Generally, each of the groups has a democratic decision-making mechanism, and distributes tasks among its members according to institutionalized rules. Particularly, development of the membership system is important to these groups. As observed by Lu, leaders and members are at a transition point to decide how to improve the quality of their activities and create more sustainable organizing systems (Lu, 2001), because they are faced with challenges to keep volunteers bound to the groups. On the one hand, internal regulations and rules would affect the interest of students in participating in activities, because young people usually do not like to be constrained by rules.[56] On the other hand, such organizational development is necessary for the groups to have a strong mobilization capacity, and keep their members even after they graduate from university.[57] In order to increase their capacity to keep members and participants, several influential groups have recently started to provide more rights to members, aiming to keep graduates interested and retaining their membership even after they graduate from university.[58]

Decision-making and implementation mechanisms are the two most important systems of SGSF's internal organization. A general committee is the highest decision-making organ of SGSF. Leader and committee members of SGSF are selected from candidates of all member groups. The implementation of decisions is completed by five departments, which are responsible for specific tasks. They include the secretary department, coordinating department, training department, external department, and information disseminating department. These parts work under the direction of a general committee. The members of these departments are also elected by democratic elections among all member groups.

With the effective management of SGSF, collective efforts of small groups were organized. Stronger impacts than that of individual groups were produced among young students and the public in Shanghai. Around Christmas 2003, 13 university student environmental groups launched an activity among Shanghai's universities, to spare trees by using fewer Christmas cards. This activity had a certain impact in the city and was reported by three newspapers.[59] In addition, SGSF helps to disseminate information and ideas on effective green activities, and promotes the exchange of ideas on management and fundraising strategies among student environmentalists.

Organizational links

Externally, as an organized force joining the forces of all member groups, SGSF has more influence than the individual groups. It has relatively good relations with other organizations, such as international organizations, other NGOs, and government agencies. Most importantly, its links with its supervising agency, the Shanghai Municipal Youth League (SMYL),[60] proves the most important one in its organizational network.

Specific member groups of SGSF are dependent on their university, because their legal status makes them secondary NGOs that work within universities. In their daily activities, this group is closely supervised by the SMYL and is directed by specific teachers, particularly on issues such as the development of targets and plans, and the means of realizing them. Although this helps the student environmentalists as young and inexperienced activists to succeed in realizing their goals, such supervision generally prevents student activists from autonomously making decisions on their programs or projects, and they act as if they are dependent on their supervising organization, advisors and teachers.[61] As a supervising institution of SGSF, SMYL functions as a bridge linking student groups with the municipal government, and has become a platform for environmental activists to interact with political authorities.[62]

As an NGO that is subordinate to SMYL, the development of SGSF depends on the attitude of SMYL. The relationship between the two is very formal. And the purpose of this link for SMYL is to incorporate green activism in its own tasks. SMYL, an organization produced in revolutionary time, could hardly continue its regulation over young people through ideological control, as in a planned Communist economy. As a way to adapt to the demands of economic development in Shanghai, it has started to loosen its political control over small groups. Links with grassroots groups were developed as a strategy of this GONGO to work with the youth in a market economy.[63] Through the bridge that SGSF makes, SMYL can continue building (indirect) close ties with the mass youth, and thus keep working as a binding institution between government and the public.

The link with the municipal government functions as an important channel for young environmental activists to articulate their ideas in the city's environmental policies. Through organizational relations with SMYL, student activists from SGSF can get updated policy information on the environmental educational policies of the municipal government, and can participate in activities such as the annual World Environment Day.[64] In recent years, SGSF has been trying to initiate advocacy activities promoted by its member groups. In 2003, SGSF wrote to the mayor and suggested that the city's preparation for the World Exposition should pay attention to ecological aspects, and advocated for channeling young environmentalists in this process.[65] The suggestion was expressed through SMYL, and was incorporated in the latter's work. Furthermore, through linking with SMYL, SGSF was able to participate in Shanghai city's environmental governance. Under the direction of SMYL, the SGSF actively joined a series of environmental mass campaigns, effectively leading its member groups.[66] In the campaign of protecting Mother River that was organized in 2003, SGSF organized volunteers from 17 universities, who presented videos, collected and analyzed water samples, and organized educational exhibition boards for the public. Such activities won SGSF and student groups recognition from the citizens. In this process, SMYL and other GONGOs also benefited from the link with SGSF to increase their own visibility and legitimacy.

In addition, SGSF and its member groups also have certain links with local ENGOs, built on the common members it shares with these other ENGOs. Through these links, the students' environmental groups can be linked with influential local groups such as GC, WCS, Roots and Shoots, and WWF Shanghai. While these double memberships promote communication between groups, they hardly play a role in facilitating cooperation and collective efforts between the counterparts. In addition, SGSF and its member groups have also built certain organizational relationship with a small number of international organizations, primarily through gaining financial support from international foundations, for instance the Global Greengrants Fund. In all, the organizational link with SMYL appears to be the most prominent organizational relationship for the SGSF and student environmental groups.

WWF Shanghai

Originally established by World Wildlife Fund, WWF Shanghai was one of the online forums of this organization that started around 2000.[67] Like WWF's web pages in other cities, WWF Shanghai's website receives loose supervision from WWF China's general office in Beijing, mainly on affairs related to the internet site. Very little formal linkage exists between the group of environmentalists in Shanghai and the Beijing office. However, a number of local residents have become member of Shanghai's digital forum and developed face-to-face interactions. In 2005, WWF Shanghai had 1000 members, although the core participants logging onto its website counted only several dozen.[68]

Members' identity

As a loosely connected group with relatively low level of organization, WWF Shanghai's maintenance depends greatly on the motivation of individual followers. It can be identified that they have developed a group identity based on their common interests. And the pleasant personal relationship that is formed among them further strengthened their integration.

WWF Shanghai's members share a common interest of watching birds, which is the main reason they meet. Several years later than Beijing and Guangzhou, Shanghai's environmentalists began bird watching in 2001.[69] Watching birds often take place in remote places out of the city, which represents one popular kind of outdoor activity among Shanghai citizens. With a decent salary, Shanghai citizens can afford the facilities needed for exploration activities. And to step out of the city from time to time to explore nature is very attractive to many city dwellers. So having fun becomes their primary goal in joining such activities, and this group of individuals consider themselves as 'happy environmentalists'.[70]

Among the members of WWF Shanghai, the development of personal relationships started only after they became followers of this loosely connected group. Computer-based communication is the main means of contact. To each insider, the virtual space that is created by the internet appears as another world, which provides a platform for communication. Through using the internet, environmental concerns, knowledge and information about watching birds are exchanged. What is more, offline activities are organized, which further lead to the development of friendship and intimate personal relationships. Aside from watching birds, members also meet irregularly at each others' homes every one or two months.[71] In this way, computer-based communication increases the ability of individuals to communicate with other like-minded activists. With face-to-face networks being formed, the network built through virtual communications is consolidated. The instrumental use of new technology in building personal contacts suggests that levels of trust among these individual citizens in Shanghai are relatively high. Common interests and ideas are enough in building close personal contacts. Relatively little idea of tradition or family values is articulated in such interactions. As one activist from WWF Shanghai said:

> 'Communicating with individuals in virtual space makes one relaxed, which differs greatly from interactions in reality. It is a pure and less complicated world than the reality. We each have a nickname in internet forum. One can behave truly. No mask has to be worn in the digital world, and I could be the real me. I enjoy it very much.'[72]

This method becomes an alternative for individuals to realize their right to express themselves freely under Shanghai's restrictive political censorship.[73] Relatively more freedom is provided than in reality, according to these activists. By using email, the activists feel more security and secrecy in organizing collective actions. These tools of communication through the internet are acknowledged by this group of participants as being more safe.[74]

Organization development

What is specific in the informal relationship formed among WWF Shanghai's members is their low level of internal organization. Yet this situation changed through the formation of another grassroots ENGO out of WWF Shanghai.

Because of the informality of this group, a very low level of internal organization exists. No individuals act as leaders. Several figures who participated in the group longer than other members are active in initiating activities and projects. Yet, because this group has no legalized status, its members did not find it necessary to become further organized. Individual members are equal, with their voluntary will as the primary principles directing informal management.

However, the informal status of WWF Shanghai affects its organizational development. Particularly, building organizational links has not been very successful. The illegal status of WWF Shanghai prevented activists from developing organizational links with other actors, such as GONGOs, ENGOs, and scientific experts.[75] Personal networks were formed only between active members of WWF Shanghai and individual media staff and scientists. Such interactions functioned as links between the volunteers and a small number of ornithological university professors. Yet, because such relationships are informal and flexible, only limited cooperation was facilitated between WWF Shanghai and the academic scholars they contacted. For instance, some scholars met with group members, and incidentally gave guest lectures to help volunteers in learning various bird species. Most of the interactions were limited to activities regarding bird protection.[76] Hardly any personal relationship was developed between volunteers of WWF Shanghai and the guest lecturers.[77] And WWF Shanghai's illegal status prevented acceptance by other institutions and actors, who feared that troubles may arise from contacts with this unregistered mass group. Registration and formalization of WWF Shanghai is greatly needed for its further development.

Under such conditions, around 2005 core members formed a new voluntary group, Shanghai Wild Birds Protection Association (SWBPA). After one year, it was registered as a secondary social organization under supervision of another GONGO—SWAPA.[78] This group has now 30 members—mainly coming from WWF Shanghai—who contribute to building the new organization. With a lack of fixed staff, SWBA is developing relatively slowly. At present, the members are planning to establish internal regulations and institutions, a membership system and decision-making regulations.[79]

This organization's establishment resulted from several developments. First, to develop into a professional association became a clear goal for the activists. After four years of activities, environmentalists in WWF Shanghai have accumulated certain experience and knowledge, and are ambitious to become a professional team committed to the protection of wild birds. Second, to be capable of mobilizing constituencies, an organization with a strict membership system is needed, which can stabilize Shanghai's bird watchers in the long term.[80] The restrictive political context also requires SWBA to be separated from the informal web-based group of WWF Shanghai. After the campaign of Jiangwan,[81] many individual followers of the group became aware that their organization was in great need of organizational structure in order to increase the group's efficiency.[82]

With a legalized identity, SWBA participates in activities oriented at birds' protection, together with professional organizations from the government and universities. It began to collaborate with Shanghai's government agencies in various kinds of bird investigations, such as the 'Winter's Water Birds Investigation', organized by the Municipal Bureau of Forestry, and the 'Investigation on Water Birds in Middle-low Part of Yangtzi', organized by the State Forestry Bureau and other institutions around the country, such as the Bird Watching Association of Hong Kong. It can be expected that SWBA will be able to build internal institutions in the near future.

The establishment of SWBA represents the development of other grassroots ENGOs in Shanghai. Internal institutionalization is very important for a group, not only for the group's survival, but also to gain social influence and build organizational collaboration and networks. Informal structures prevent WWF Shanghai from being recognized by other organizations, and thus constrain the formation of organizational development in the long term.

Campaign protesting against the construction of New Jiangwan Town

This campaign was the first collective effort organized to oppose environmental policy in Shanghai. Under the restrictive rule of the municipal government, WWF Shanghai's activists failed to effectively mobilize resources. Nevertheless, the activists managed to articulate environmental interests, and to some extent affected local policy implementation.

Introduction to Jiangwan marshland

Built in 1922, Jiangwan Airport was the first military airport in China. It is one of the historical sites of the Sino-Japanese War, where the Nationalists of Kuomin Tang used to fight. After 1949, it was used as an air base for the Chinese People's Liberation Army. Two decades later, this area turned into Jiangwan marshland, covering an area of more than 8 square kilometers, and accounts more than 1 per cent of the whole city (Map 6.1). Because of its rich water resources and the lack of human disturbance, the area evolved into a complete ecological network and a haven for wildlife. It is said that more than 280 wild plant species and nearly 40 freshwater plants are located in the Jiangwan marshland, of which many are rarely seen. Almost 90 per cent of summer birds in the city breed in this area, according to environmental activists' estimation.

After the land ownership was turned from the Chinese Liberation Army to Shanghai municipal government in 1997, the latter planned to develop it into a high-class sub-city center in the northeast of the city.[83] Shanghai New Jiangwan City Investment Enterprise (SNJI), a state-owned enterprise that focused on Jiangwan's development, was established. The marshland was to be constructed into New Jiangwan Town, one of the four sub-centers of the municipality, as a modern residential area integrating businesses, logistics, cultural facilities, and residential communities, accommodating a population of about 80,000.

Yet this project was acknowledged to the public only at the end of 2001, when New Jiangwan Town began to be constructed. During the formation of this policy, no EIA reports were conducted, neither was the public's opinion consulted on this policy. Under such circumstances, environmental activists began to protest against the marshland's destruction.[84] A campaign by environmentalists from WWF Shanghai was developed, which can be divided into two stages. The group acted as a leader mainly in the first stage, while its impact began to decline in the second phase.

First stage

WWF Shanghai was the main actor coordinating the campaign against the construction of Jiangwan Town in the first stage. It organized a limited number of events, which illustrated its central role among actors in this campaign.

In this phase, environmentalists faced difficulties in effectively mobilizing the resources they needed. Prevented by its informal organization and illegal status, WWF Shanghai could hardly build a professional scientific team to explore the value of the marshland. Although investigations on Jiangwan marshland were organized using 200 WWF Shanghai's followers, interested citizens and a small number of academic scholars, no concrete scientific report was put together to prove the biological and ecological value of the marshland. The environmentalists gained their knowledge of the marshland primarily

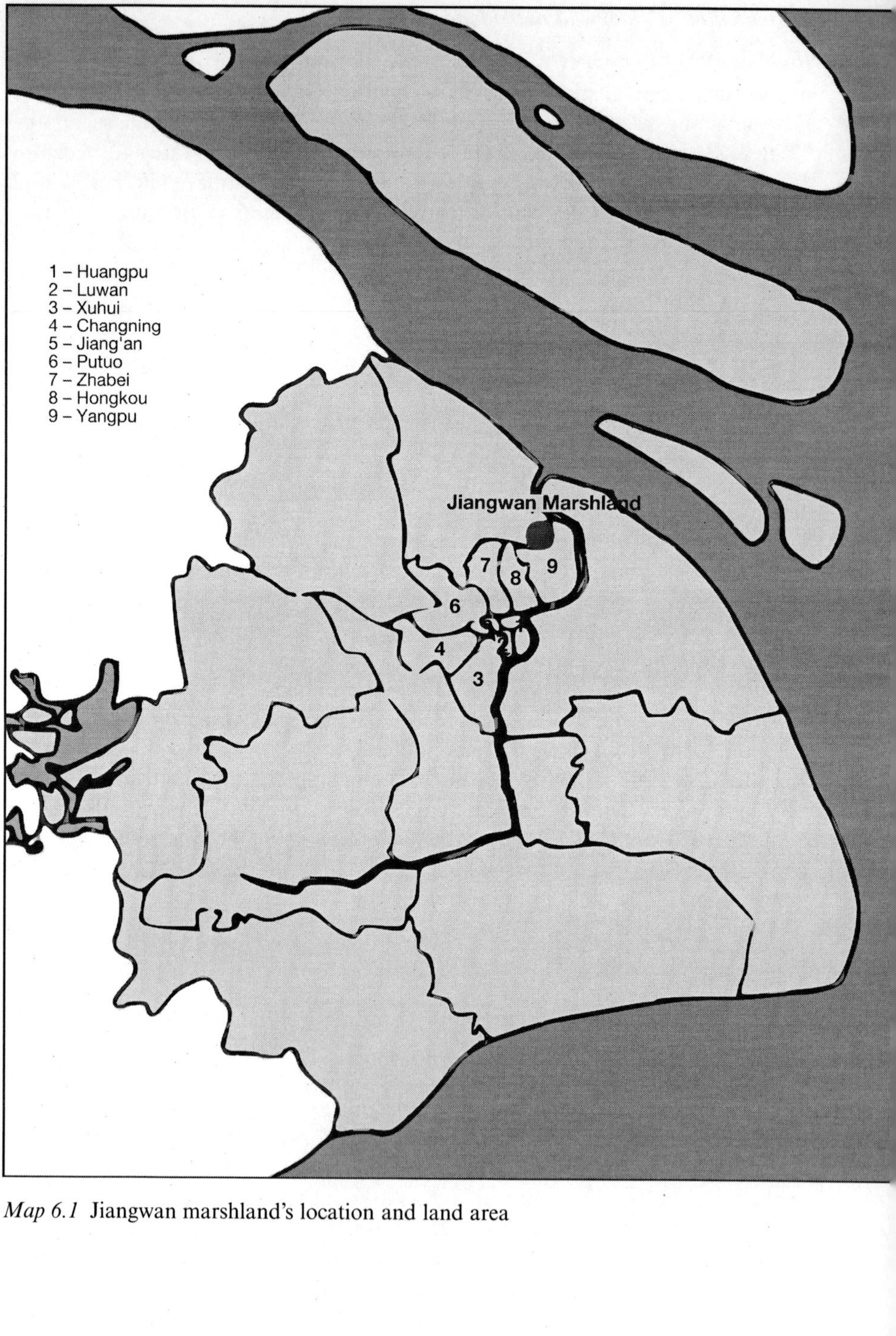

Map 6.1 Jiangwan marshland's location and land area

through observation and investigation. One activist, Mr Li, is very familiar with the evolution of Jiangwan after several years of exploring this area in his spare time. His knowledge became the main source for developing a report on the ecology of the marshland.[85] But this report, produced by environmentalists, seriously lacked validity.

In addition, what also weakened the environmentalist force was that very few social groups joined this campaign, mainly because of the issue's political sensitiveness. For ENGOs with which WWF Shanghai had links, such as GC and students' groups, opposing the government on this issue might pose serious risks to their status. Being a grassroots group without legal identity, GC's leaders were hesitant to take any action. And under supervision of university teachers, students in SGSF were constantly reminded not to take any action.[86] As a consequence, no integrated alliance could be organized in this campaign. Only environmental activists in web-based groups, such as Shanghai Nomad Club—a traveling group—and Common Wealth Management Project—an online group committed to provide education and services—were mobilized to join the investigation and sign the petition.[87] Such links were built through personal networks among the common members that WWF and several web-based groups have.[88] Yet, because of the loose organization of both counterparts, limited collective efforts were made.

Lacking experience and careful plans, building public opinion constituted the primary strategy that environmentalists utilized to attract political recognition through the campaign. This was mainly organized through conventional mass media and the internet, and local journalists were contacted. Yet, under pressure from the Yangpu District Government, only a limited number of preliminary news reports on the marshland were completed and published.[89] Soon, further reports on this issue were banned. Computer-based communication then remained as the only tool for releasing news and constructing public consensus. Based on the electronic resources of members, such a strategy was still quite influential among the public who had access to the internet. Information on the destruction of the marshland was distributed to a number of online forums of web-based groups, including Shanghai Nomad Club and Common Wealth Management Project. What is more, a special website dedicated to marshland protection, named 'The Event of Jiangwan', was developed, on which comprehensive information and pictures were placed. In order to increase the number of readers, this website was also linked to the official website of Shanghai People's Congress, which was holding conference in 2002. In one week, visitors to this site reached 3000, and the number continued to grow. Although the government blocked this site after one month, it was efficient in spreading the news up to that moment. Seen from the increasing number of visitors to the website, it was computer-based communication that increased the dissemination of information. In early 2002, through interaction on the internet, public opinion was influenced towards protecting Jiangwan, as the last marshland of Shanghai.

Later on, meetings were organized among environmentalists for further dissemination of information on this issue. In order to broaden the campaign's influence, botanists, ecologists, ornithologists, and environmental scientists were contacted and invited to participate in meetings.[90] A meeting at Huadong Normal University was held, with around 70–80 people, including scholars, experts, and activists from WWF Shanghai. On this occasion, the values of Jiangwan marshland were discussed, which led to the signing of a petition, both off- and online versions. As a result, around 1000–1500 signatures were collected. Because several scholars served as members of the Shanghai Municipal People's Congress (SMPC), this petition was inserted into the formal policy process by submitting it to the Tenth SMPC of Shanghai in 2002. There it was suggested that the municipal construction department should reconsider the original plan of city construction, and establish a nature reserve around Jiangwan marshland area.[91]

Following the SMPC meeting, the municipal government halted construction. Shanghai Construction Commission was ordered to reinvestigate this issue. For the first time, the public, led by activists, scholars, and political representatives, had successfully influenced local government policy. Although the district government took strong measures to repress publications on the campaign in the mass media, the scientific support appeared to be an important resource affecting local policies. This demonstrates that the public could influence this process through institutionalized policy channels.

Second stage

In the second stage of the campaign, starting in late 2002, Yangpu District Government, supported by the City Planning and Construction Bureau of the municipal government, adopted repressive actions to weaken the coalition of environmentalists and scientists. The collaboration between WWF Shanghai and its partners in the movement was counteracted through interventions of political authorities. In addition, government agencies took the chance to initiate a green 'washing' of the project via its control of the majority of mass media in Shanghai.

A split in the coalition was started by the separation of individual scientists, who were warned by their direct authorities in universities and institutes.[92] As the environmentalists had no scientific proof in their report on the Jiangwan marshland, and were not formally supported by any professional academic organizations, their influence started to fade. What is more, they did not try to seek collaboration with actors in other provinces to broaden their influence outside Shanghai. Aside from the scientific experts who originally collaborated with the environmentalists, local journalists received similar reminders from the political authorities. Not only reporters who worked for the Party's newspapers, but also those less affiliated to Party publications, were prohibited from continuing to report on this issue.[93] Losing partners in sustaining and expanding public opinion on the protection of the marshland, the impact of environmentalists on this issue became largely weakened by late 2002.

At the same time, a new public opinion, constructed and promoted by the municipal planning agencies, was established and dominated policy-making. This process was largely controlled by SNJI and the City Planning Department of Yangpu District Government. When launching the EIA, the authorities managed to form a positive consensus on the construction in the marshland, the result of which to a large extent contradicted the evaluations given by the scholars and experts in their report to SMPC.[94] Scholars who originally opposed this issue were left outside the EIA process.[95] Thus this EIA was accused by environmentalists of lacking transparency. Along with the construction of the EIA report, a number of articles were released in official journals and newspapers, which generally supported the exploitation of the marshland, strongly pushed by the authorities.[96] As a result, a revised construction plan was brought out. In the new plan, only an area of around 200 square meters was preserved.[97]

The dynamics of the campaign illustrates that in the coercive political atmosphere of Shanghai, very little space is available for organizing collective campaigns. Yet the characteristics of the informal ENGO also contributed to the limited role it played in the end. Although highly committed to this campaign, individual activists were too inexperienced to develop successful protest against a repressive government. They could garner only limited sources of support, which resulted in cooperation only among local actors. No other national ENGOs or international actors were mobilized. With very few national media contacted, the public report on this issue was conducted mainly by local reporters. Within a short time, they were disconnected from the environmentalists through political authority intervention.

A low level of coordination can be recognized in the campaign, which resulted from a lack of collective identity among movement actors. Intentions and interests on this campaign varied among movement actors. Individual volunteers of the WWF Shanghai group were strongly influenced by the charisma of Mr Li, who joined the group for initiating the campaign.[98] This person's persistence motivated and encouraged volunteers to commit themselves to the campaign. As to the academic scholars, their intentions in participating varied. Although they generally opposed constructing the new estate, their ideas on protection differed, for instance between botanists and ecologists, due to differences in scientific points of view.[99] The diversion of views to some degree mitigated their involvement with environmentalists. Without a clear and strong collective identity among the environmentalists and scientists, the movement actors produced few successful collective efforts.

With the short duration of protest and limited revision of the original plan, we can conclude that the social leverage of this campaign was very limited. Nevertheless, individual activists articulated their opinions through collective efforts, and expressed their demands for transparent policy-making. In the enclosed political atmosphere of Shanghai, this is one of the few cases where environmental policies have been influenced by civil society groups.

Conclusion

In general, illustrated by this case, limited space is available for coordinated efforts from the ENGOs under the restrictive environment of Shanghai. Two reasons exist. First, demonstrated by the campaign to mobilize citizens to protect the Jiangwan marshland, political authorities are very cautious with respect to the emergence of civil society organizations and collectively organized efforts to oppose government policies. Local environmental activism thus develops slowly, compared with similar sized cities of the same economic development level. And second, governmental agencies greatly encourage GONGOs and market actors in environmental governance. They rely heavily on GONGOs to coordinate the public in such environmental governance processes. This leaves grassroots ENGOs and environmentalists few resources and channels through which to get involved.

Despite the repressive environment, growing green activism can be recognized in Shanghai, which shows particular characteristics of environmental organizations, primarily with respect to their internal organizational development and the strategies they adopt. Internally, organizations that have a high level of institutionalization and rich organizational resources play an important role among local ENGOs. In comparison, building formal internal organizations is a common theme among all three ENGOs. As the oldest group, GC has the highest level of formality, which brings effective internal coordination and leads to strong action capacity in its activities. Along with its further development, WWF Shanghai began to take the initiative to establish internal institutions. This indicates that the longer Shanghai's ENGOs exist and the more they become engaged in public activities, the more they formalize their internal organization. This tendency of Shanghai's ENGOs is not unlike what is reported on environmental organizations in Western societies (Van der Heijden,

Table 6.2 Characteristics of members and grassroots organizations

	WWF Shanghai	*SGSF*	*GC*
Age	88.5% 22–35 Average 29	18–25 Average 23	Average 29*
Income (Yuan)	<2500, 32.8% 2500–5000, 44.3% 5000–8000, 16.3% >8000, 8.1%	None	Moderate
Occupation	Private company employees Professionals and a small number of university students (10.7%)	University student	Professionals, intellectuals and a small number of university students
Level of education	>91.8% of members received university degree or above	University degree or above	>90% of members receive university degree or above

Notes: *Data estimated by members of GC: the number counted by average age of volunteers and committee leaders. The average age of volunteers is estimated at 28, and committee leaders are aged 30 on average.

1997). What is more, externally, Shanghai ENGOs show a relatively weak reliance on individual networks. WWF Shanghai and GC rely marginally on personal networks composed of common members. But all ENGOs show little sign that close personal contacts are crucial as links between groups and with government agencies. The specific social circumstances have not led to a strong reliance on, and utilization of, personal networks.

Two contextual reasons contribute to this characteristic of Shanghai ENGOs. First, the local culture is less encouraging for the utilization of personal networks, which reduce the possibility that the strategy of personal connections is strongly adopted. In Shanghai city, utilizing *guanxi* is not a widely prevailing and accepted behavior. The Shanghai 'subculture' favors the establishment of rules and formal systems, as fostered by the development of business and finance in Shanghai's contemporary history. And this inclination to follow formal established institutions is further strengthened by its restrictive political authorities.

Second, the formalization development of ENGOs is also related to the characteristics of environmentalists in these three groups, which show a large degree of homogeneity (Table 6.2). A high proportion of them are young and well educated.[100] Professionals, particularly managerial personnel, account for a high percentage of core volunteers. This group of people are inclined to rely on formal mechanisms to regulate their work and activities. To them, constructing principles and institutions is the most effective way to conduct teamwork.[101]

When specifically comparing the ENGOs in this study, it is found that the social influence ENGOs have is related to the resources they can mobilize

Table 6.3 Comparison of resources mobilized by ENGOs

			WWF Shanghai	
	GC	*SGSF*	*First stage of campaign*	*Second stage of campaign*
Government access	ZBEPB; ZBYL; Zhijiang Xilu sub-district	SMYL	None	None
NGO network	Local NGO; national NGO; rich connection with INGOs	Local ENGO; a small number of INGOs	Web-based NGOs; very few INGOs	Web-based NGOs
Science and expertise	Knowledge from INGO	None	Informal links with scholars and scientific experts	None
Media access	Little	Little	Informal links with local media	None
Financial sources	Sufficient	Relatively sufficient	Limited	Limited
Membership	Large number	Medium size	Large number	Small number

(Table 6.3). For instance, GC possesses abundant resources, and acts as a broker among the environmental groups. On the other hand, WWF Shanghai, after political authorities had intimidated the media and scientific experts with whom the environmentalists cooperated, experienced limitations in resources in the second phase of the campaign, and thus the group ceased to act as broker in the campaign. This case demonstrates the importance of the availability of resources in mobilizing social movements, analyzed by resource mobilization theory (McCarthy *et al.*, 1973, 1977). With limited means available, environmentalists will not easily be able to achieve strong influence and realize their goals.

7 Comparison and conclusions

Introduction

During the past decade, voluntary environmental groups and environmental activism have emerged in China. Characterized by their commitment to improve environmental quality and environmental governance, environmental NGOs (ENGOs) and social organizations are becoming the building blocks of a growing environmental movement in China. With increasing environmental degradation and resource depletion, and the inadequacy of the state in environmental governance, the emergence and organization of environmentally concerned citizens should come as no surprise. But these environmental groups still look very different from what we are familiar with in Western Europe and North America. In comparison to the environmental movement in Western Europe and North America, Chinese ENGOs are limited in their capacities, as they lack resources to act independently and influentially, such as financial support, a knowledge and information infrastructure, social recognition, and a large constituency; and they are restricted in their freedom to act under the current political and legal system in China.

In order to understand and study the functioning and development of environmental groups in China, one has to analyze the interactions between these groups and their external social and political environment. In China, the authoritarian state leaves very little space for self-organized social groups and their involvement in politics and policy-making. It poses various obstacles for collective efforts made by environmental activists and, in order to understand the development and functioning of social movement organizations and their specific (Chinese) characteristics, one needs to be aware of these national specifics. In this social context, a social movement can be defined as 'networks of informal interactions between a plurality of individuals, groups and/or organizations on the basis of shared collective identities' (Diani, 1992).

In China, a concept of civil society organizations or social movement organizations not associated to the ruling party is quite new. The development of ENGOs can be traced back to the early 1990s, when Chinese environmentalists were greatly enlightened by the Rio Declaration, following the 1992 UN Conference on Environment and Development. Later, in 1995, female voluntary groups revealed their significance in co-organizing the World Women's Conference in Beijing. These two events encouraged and inspired many Chinese environmentalists in organizing their own voluntary groups and organizations, resulting in the first boom of Chinese NGOs. While data are mostly unreliable and vary widely, as discussed in Chapter 1 (also due to the inexact definition of ENGOs in China), some believe that until 2005 approximately 2000 environmental groups were officially registered as NGOs, with probably as many registered as for-profit business entities or not registered at all (Economy, 2005).

The significant number of unregistered environmental organizations reflects the government's suspicious attitude toward NGOs in general. Fearing political instabilities that may occur through the booming of NGOs, the state administration constrains the growth, form, functioning, and strategies of civil groups. Under the state's repressive control, very few ENGOs are inclined toward adversarial or confrontational strategies and goals which might contradict government policies. Coordinated actions that are not completely in line with, or fully supported by, the government find it difficult to procure support. And special tactics are needed in protesting and campaigning activities. Most ENGOs prefer to engage in raising environmental awareness among citizens and officials, in disseminating information to the wider public, and in clean-up actions in streets and parks; and they refrain from adversarial, confrontational, and media-based campaigns. In general, groups committed to environmental education compose the majority of NGOs.

In contemporary China, opportunities and legal obligations concerning public participation are very limited. While several environmental laws require the government to inform the public, few allow the public to be involved in policy-making processes and decisions. Only in 2003, the newly promulgated Environmental Impact Assessment Act confirms the right to public participation in environmental policies. This Act requires that certain projects with significant potential environmental impact should involve consultations with the public, by publicizing information regarding the project and its environmental effects, or holding public meetings. Unfortunately, in practice this law is not always implemented strictly (Beach *et al.*, 2006). In reality, the public has very limited institutional access to policy processes, which remains very much a promise and challenge for the future (Buckley, 2006; Chung, 2006).

Under such conditions, one can expect that environmental groups have very limited formal, institutionalized channels through which to have an impact on environmental policies and decisions. Informal channels such as individual networks and personal ties can then become an alternative route of access to policy-making and a way to organize collective actions in Chinese society. The complex of personal connections or *guanxi* can be essential in environmental activism, and has been the prime object of this research. To what extent, and how, are individual connections and relations among friends, relatives, colleagues, neighbors and so forth building blocks in the construction of the Chinese environmental movement? Pre-existing personal ties from previous social movements and collective actions are often well preserved, for example those from the 1989 Student Movement at Tiananmen Square. But the question is whether the role of these personal networks in environmental activism relate (only) to the specific political situation of contemporary China. Numerous authors have argued and illustrated the wide permeation of such networks throughout Chinese culture and society, also in earlier time periods. In addition, one could explain the importance of these personal networks in environmental activism by the developmental stage of the Chinese environmental movement, which is still in an early phase of professionalization, formalization, and institutionalization.

Irrespective of the causal relations behind the prevalence of personal contacts and networks in the Chinese environmental movement, focusing on these individual or personal networks (the two concepts are used alternatively) is crucial to understanding the current functioning and development of the still immature Chinese ENGOs in a situation where the strategies, resources, and organizational schemes of Western NGOs are predominantly absent. In the individual networks, we find many elements of the dynamics currently used by environmental activists to protect and preserve the environment.

This study explored the role of personal networks in the contemporary environmental movement in China. Case study methodology was used to examine how individual networks have an impact and function in the environmental movement in four urban areas: Panjin, Beijing, Shanghai, and Xiangfan. In these cities, the environmental movement has been analyzed, and sub-cases of ENGOs and their environmental campaigns examined.

In this final chapter, the empirical material from these four case studies is compared, the research questions answered, and further conclusions and research directions formulated. Following this introduction, the chapter is organized into three main sections. In the following section, the results of case studies are compared and analyzed with respect to the utilization of personal networks. Part of this comparative analysis is to investigate how social circumstances in each city affect the use of personal networks by environmentalists. In the next section, the research questions as formed in the Introduction and Chapter 2 are answered and conclusions are drawn. In the final section, a future research agenda is formulated based on this research.

Comparative analysis of case studies

A methodological introduction

The investigation into ENGOs in the four case study cities has illustrated that adopting *guanxi* networks constitutes an important feature in Chinese green activism across different geographical regions of the country. This study has chosen four case study cities to study local green activism, to provide sufficient diversity with respect to economic level, the openness of local political authority, and the prevailing local culture. Before analyzing and comparing how Chinese environmentalists utilize personal networks in these four regions,

I will briefly reflect on the diversity of the case studies, by reviewing the relevant contextual aspects of each case.

Panjin is not a very wealthy city, with a relatively short history. It is experiencing fast economic development and modernization under the direction of a relatively closed political system. The city lacks a prominent specific local culture, but has been greatly influenced by the northern provinces during its development. As the northern provinces were for a long time dominated and influenced by minorities and nomad nationalities, and experienced constant shifts of rulers and frequent migration, this area—and thus also Panjin—lacks stable and lasting cultural traditions. The local subculture that has grown out of these influences is, to a large extent, built around the utilization of *guanxi* and personal ties in politics and economic networks. The repressive political system contributes to this strong reliance on personal networks in many sectors. This phenomenon is reflected in the emergence and development of green activism, where gaining resources and support from local political authorities, in particular, is mediated via personal networks.

Xiangfan, a poor metropolis with a relatively long history compared with other Chinese cities, has developed under the influence of cultural traditions and conventions. Although in transition, the city still witnesses a strong reliance on a planned economy in its economic growth model, more so than other Chinese cities of comparable size. Located in middle China, this city sees relatively slow economic development, compared with those along the east coast. Its relatively closed political system seems inefficient in promoting an economic transition to establish a well functioning market economy locally. Influenced by local culture, citizens greatly value family and personal connections. Building personal relationships and using them practically in economic and political engagements is very prevalent in Xiangfan. In the growth of green activism, personal networks are essential in mobilizing common identities and organizing collective actions.

Beijing, the capital of China, has high levels of economic development and economic growth, and a relatively open political system, also caused by its embeddedness in international networks. The city has over 700 years of history. Its subculture greatly values interpersonal relationships, but through its international relations in politics, economy, science, and increasingly also civil society, the professionalization and formalization of organizations are progressing. This high international visibility works in various ways on local politics and civil society. Repressions are usually less severe, but highly controversial protests are strictly controlled, especially since the Student Movement of 1989. Influenced by local culture, the practice of using personal connections is prevalent to some extent, but is used especially in the organization of national campaigns.

Shanghai is the second largest metropolitan city in China, and has experienced fast economic development. It is the model for the east coast cities, all of which have thriving development and have witnessed a rapid transformation towards a market-oriented growth model. Shanghai has been an important city in Chinese contemporary history. As a port city, it represents a mixture of both Chinese and Western cultural values. Its political and economic systems rely less strongly on personal networks than do other Chinese cities. The local culture equally witnesses less intense interpersonal relationships. Its political system is very repressive, which clearly distinguishes this metropolis from Beijing.

Together, these four cities encompass a significant part of the diversity of China with respect to economic development, local cultural systems and traditions, and political openness and repression. But we should also recognize that important loadings of these contextual factors are not included in our selection of case studies. No major city from one of the western provinces was included in our sample (which would probably show differences in economic development and local culture); and all cities were of considerable size (with over 1,000,000 inhabitants), leaving the more rural developments in towns and villages out of our analysis (which might affect the relevance of personal networks and of the degree of repressiveness of political systems). Further, the types of ENGO that are found are primarily those committed to environmental education (which might not cover all possible strategies that Chinese environmental organizations adopt). Consequently, we cannot generalize the case study findings regarding the functioning of green activism and their use of local networks to the entire Chinese nation. While the sample of case studies is sufficiently diverse on relevant contextual factors to be able to touch on differences in the role of personal networks in the functioning of the (emerging) Chinese environmental movement, we cannot claim that with these case studies we have covered the diversity of China. And we certainly cannot claim that the findings on Chinese environmental organizations are representative for the entire 2000 (or more) environmental movement organizations in contemporary China.

Consequently, the nature of our analysis will essentially be to analyze when (under what kind of contextual conditions) and how personal networks are (not) used and applied to protect the environment in major Chinese cities.

Functions of personal networks in environmental movement

In exploring how personal networks are used in Chinese environmental movement organizations, I first compare to what degree environmental activists from the four cities rely on this mechanism; then I specify what functions these personal networks fulfill for ENGOs; and finally I focus on conditions that affect the utilization of these *guanxi*.

Although the phenomenon of using *guanxi* networks is widely recognized in Chinese green activism, the four case studies illustrate variation in the utilization of these personal networks in Chinese ENGOs. Table 7.1 summarizes the different degrees to which NGOs rely on personal networks. In Panjin, green activism relies heavily on *guanxi* networks. In Xiangfan, personal networks proved to be important in the development of Green Han River (GH)

Table 7.1 Degree of reliance on personal networks

	Extent of reliance on personal networks
ASGP (Panjin)	Very strong
GH (Xiangfan)	Strong
GEV (Beijing)	Weak
GC, SGSF and WWF Shanghai (Shanghai)	Very weak

and the initiation of its activities. Consequently, it was found that the dependency of this organization on the personal relationships of its leader makes it very vulnerable. In Shanghai, personal networks are developed through common membership of NGOs without one member acting as a definite leader. A small number of common members act as links bridging different ENGOs. Because the identity of these individuals together defines the profile of the organization they belong to, the personal links that are formed also bear organizational characteristics. The Shanghai NGOs are less likely to be formed through, and draw upon, personal relationships. So when compared according to the extent of reliance on personal networks, Panjin is the city where the degree of reliance on personal networks is very strong, and Shanghai is the case where the lowest degree of reliance on personal networks could be identified in this study. The use of personal networks in Xiangfan and Beijing is in between these cases, with Xiangfan being more dependent on personal networks than Beijing.

The reliance on personal networks of NGOs varies in the four cities, and the functions these networks fulfill for the NGOs also differ significantly. When we focus on the functions of personal networks in Chinese environmentalism, we find they are effective in three dimensions: affecting political authorities, forming environmentalism as a collective identity, and mobilizing coordinated actions. Table 7.2 summarizes the functions of personal networks for the various NGOs in the four case study cities within these three dimensions. Below we elaborate and compare each of the three dimensions.

Affect political authorities

The case studies illustrate that personal networks are used in the interaction between ENGOs and the political authorities, through which various political resources are garnered, such as environmental information and data, and the application of political authority needed for mobilizing environmentalism.

First, by using personal networks, NGOs are able to acquire useful information on environmental problems, protection, and governance. The availability of data on environmental quality and emissions, and environmental policy- and law-making, is very limited, because the Chinese political authority monopolizes, to a large extent, environmental information as well as the environmental policy making process, without providing room for other organizations or parties to participate. Utilizing personal networks thus helps ordinary

Table 7.2 Functions of personal networks in environmental campaigns in urban China

	Panjin (northern China)	*Beijing*	*Xiangfan (middle China)*	*Shanghai*
Access to political authorities	Gain political authority	Withdraw policy information	Withdraw policy information and gain political authority	None
Form environmentalism as collection identity	Constructed by exchanging personal interests	Environmental affection articulated and disseminated by personal networks	Personal network strengthens existing environmental affection with values of collectiveness	Personal network has little function in forming local environmentalism
Mobilize coordinated actions	Links built among top officials in local government and media; produce social leverage through building local public consensus	Links built among MEP, media; scientific experts and local NGOs; produce social leverage through building national consensus	Links built among EPB; XCPPCC and national ENGO, social leverage produced through building local public consensus	Links built among media, scientific scholars and local web-based groups

citizens and ENGOs to get in touch with political agencies and access information, including the making of environmental policies. Evident in the ENGOs that were analyzed in the cases of Beijing and Xiangfan, through personal ties that the leaders of ENGOs have with political elites, updated information about policy, projects, and environmental data were provided to environmentalists. This was less evident in the cases of Shanghai and in Panjin.

Second, another resource mobilized through personal networks with political authorities is various types of political protection and assistance by these authorities. As noted, Chinese citizens and civil society organizations have a tendency to regard the government as the undisputed authority for all kinds of public affairs, and are unlikely to oppose it. In this situation, ENGOs, as representatives of the public, have to gain access to authority and protection by governmental agencies before they can produce social leverage. This is often already in place when registering NGOs, and continues when ENGOs carry out various environmental protection activities through which they criticize current (economic, political, or societal) developments. Linked by personal ties, political officials could endow key figures of ENGOs with certain resources, authority, legitimation, and protection in carrying out these activities. Such interaction is generally conducted based on face-to-face contact, personal trust, and exchanges of mutual advantage, and could scarcely be replaced by organizational networks.

Facilitating environmentalism

In different cities, where local cultural, economic, and political conditions vary, different types of environmentalism are formed, as illustrated by our case studies. The empirical findings of this study demonstrate that at the micro-level of the environmental movement, it is through personal networks that ideas are articulated and communicated and collective identities formed. In Panjin, Xiangfan, and Beijing, this formation of identity and communication through personal networks primarily came about through face-to-face contacts; while in Shanghai, electronic communications played a major role.

In the case of Panjin, a *guanxi* network was delicately developed by the leader of the Association of Saunders s Gull Protection (ASGP). It was built on a small scale in the city, still functions on that level, and appears to be vulnerable and instable. Only a relatively weak collective identity was formed via *guanxi* in this NGO. The leader plays a crucial role in linking the individual members and sustaining group solidarity, making his position in the local environmental network irreplaceable.

In Xiangfan, the collective identity that is formed among environmentalists shows citizens' passion for their rich local ecological environment. Ms Yun, leader of GH, also strengthened this passion with ideas that value collectiveness in local culture. In this city, the inhabitants, young or old, still adhere to an ideology promoted by the Communist Party that requires obligations and self-sacrifice, which has expanded to environmentalism. Built on trust, familiarity, and personal charisma, the leader of the GH successfully built on this cultural heritage a collective identity for the NGO. In this way, local environmental activists show strong and committed activism in improving environmental reforms.

In the case of Beijing, personal networks have been instrumental in reinforcing existing individuals' consciousness and articulating a movement identity. In this city, where cultural influences tend to be strong, individual citizens, particularly the social elite, share the solidarity of shouldering social responsibility for the whole society. Through Ms Wang's personal network, this cultural background was combined with passion for the environment. And she utilizes her *guanxi* network to disseminate ideas extensively. In this way, the NGO succeeded in mobilizing new individual participants.

In comparison, in Shanghai the role of personal networks is of less use in the construction of collective identity. As citizens of a cosmopolitan city, local inhabitants are active in sharing ideas and values and approaching people of a similar outlook. In these circumstances, collective identity appears to be established relatively easily through individuals' self-expression and the joining of like-minded people in real and virtual locations. Communication among movement players was, compared with the other locations, strongly computer-based and did not rely only, or even primarily, on face-to-face interactions. The function of personal networks thus appears to be less prominent in building collective identities in Shanghai than in other cases. Common interests and like-minded people grouped together and co-constructed the collective identity of an NGO, rather then using one central leader's personal networks to construct a common identity of environmentalism.

The content of environmentalism among these case study NGOs varies in three aspects: the geographical scale of identity formation; the issues of concern; and the underlying ideologies. First, Chinese environmentalism focuses primarily on local environmental issues. As found in the cases of Xiangfan, Panjin, and Shanghai, the environmental identities are based on issues of local geography, where local environmental issues mix with city or regional identities. Only in Beijing, the capital, are nationwide environmental issues relevant in formation of a green identity. As introduced in Chapter 1, citizens are likely to be more concerned about issues that relate directly to their local living environment, where issues are easily shared and disseminated, and become part of a common identity. This is related in part to the heavy reliance on face-to-face communications through personal networks, rather than mass media communications and the dominance of an organizational network. Second, in the issues of interests identified in the case studies, protecting nature and wild animals are well represented. While initially these might be considered less politicized and have fewer chances of conflicting with the dominant development models supported by the state and Party, in practice this is not always true. Very few issues are raised and formulated in ways that (may) challenge the ruling Party and its legitimacy. This also explains why contemporary green activism is accommodated relatively easily by the state, although the example of the dam campaign proves that we might be on the eve of a major change. Thirdly, the ideology of Chinese environmentalists has been influenced by local cultural backgrounds. In both Beijing and Xiangfan, local culture strongly influenced Chinese environmental activists. Values such as responsibilities and collectiveness became important elements in local green activism. This explains why ENGOs are able to mobilize individual 'members' successfully, even when their membership systems are poorly developed.

Mobilizing coordinated actions

Personal networks are also used for forming coordinated actions, which is important in the restrictive political climate of China. This is also what distinguishes Chinese environmentalism from its Western counterparts, where institutionalised systems and organizational networks are crucial elements of coordinated actions, rather than personal networks. Chinese environmentalists manage to generate a certain social leverage through their coordinated actions and activities, and personal networks usually play a significant role in this. Two interrelated dimensions are relevant here: through personal networks, linkages among actors from different fields and sectors are established, and consensus is built as a preferential strategy for collective action in contemporary China.

First, Chinese ENGOs activate personal networks in linking other sectors of society to heighten pressure, organize activities, and build up consensus in support of their ideas and goals. The case studies illustrate that personal networks among Chinese ENGOs prove especially useful and important in incorporating the media and scientists in mobilization and collective action processes.

The media are a very important player in the Chinese environmental movement. As seen in all four cases, newspaper staff collaborated in building public consensus on the objectives and ideas of the ENGOs. Whereas efforts to build up public consensus through the media take a long time, the social impact can be powerful. In the case of Beijing, for approximately one year media reports were published in both national and local newspapers on a dam construction dispute and the negative ecological effects that could result from it. In the end, this media campaign had a distinctive influence. To achieve this, personal contacts with journalists and media officials should be well established. The presence of reporters in many of the environmental movement networks in our case studies illustrates this.

The reason that the media play an important part in the contemporary Chinese environmental movement is related to two major developments. Throughout the past decade, more and more media have obtained growing independence from the Party state. To some extent they had to become increasingly economically independent as state subsidies diminished, while at the same time state control relaxed somewhat on those issues that were not of high political importance.[1] In this process, the public's growing environmental awareness could be strongly reflected through increased coverage on television and radio and in newspapers. Secondly, in recent years the central government, particularly led by the Ministry of Environmental Protection (MEP), began to strengthen environmental education at the local level. In order to realize the central government's policy, many more news reports and programs are required to help educate citizens on environmental protection. This has provided more possibilities for environmental media reporting at various levels and through different media.

Scientists are another strategic partner for the Chinese environmental movement, not unlike many other countries. Many personal networks of leading environmental activists include scientists. As indicated in the case of Beijing, scientists provide scientific proof for the ENGOs, which often lack expertise, scientific perspectives, and scientific authority. The participation of scientists is enabled and motivated, as they have more freedom of expression, criticism and information access under the authoritarian state than environmentalists. As argued by Cao & Suttmeier (2001), with the development of a market economy and the transition of the new political leaders, scientists and scholars began to be better supported by the ruling authority.

Through using the personal networks of leading figures in the ENGOs, movement actors are not constrained by organizational boundaries. In comparison with formal mechanisms and organizational networks, the strategy of using *guanxi* connections appears to be flexible and effective under Chinese conditions. In Panjin, *guanxi* established between the leader of the ASGP and his colleague journalists, facilitated media consensus in the local media, which successfully improved local policy on the protection of gulls. In Xiangfan, where the political system is relatively closed, the leader's *guanxi* network with government officials enabled access to the political system from outside. It is difficult to see how, in these situations, personal networks could be replaced by organizational networks.

Secondly, the case studies demonstrate that through the *guanxi* networks of ENGO leaders, coordinated actions were realized in building public consensus and thereby producing political power in the Chinese policy-making processes. Consensus-building is essential for agendas that are not given sufficient priority in the state bureaucracy, in order to gain clout with the more powerful units in the fragmented Chinese bureaucracy (Lieberthal & Lampton, 1992). By adopting this consensus-building strategy, the Chinese environmental movement manages to gain influence in the policy and decision-making processes that are consequential for the environment.

In environmental governance, the strategy of establishing public consensus is recognized in each of the four cases. Of the four cases, two established national consensus, while the other two focused primarily on local consensus-building. The more renowned the affiliated media, and the more media channels involved, the more influential a public environmental campaign could be. In the Xiangfan campaign, through personal networks, news reports were posted on the international media circuit, websites and newspapers protesting against river pollution, which activated more prominent and powerful forces than the campaigns in Shanghai, which used only local newspapers for mobilization of collective action. In Beijing, for 12 months a dispute on dam construction was publicized in both national and local newspapers and journals. In Xiangfan, provincial media were used first to build public consensus. In both cases, government environmental policies received considerable pressure.

In this process, personal networks serve to facilitate communication and organize public consensus. In the case of Beijing, the leader has wide personal connections with reporters, and could successfully apply pressure on politicians at the national level. Green Environmental Volunteers (GEV) thus played a leading and agenda-setting role in the campaign protecting the Nu River.

Factors influencing the adoption of personal networks

In this section, I assess the influence of contextual factors on the utilization of personal networks: economic development, openness of the political system, and local culture (Table 7.3). But as reported in Chapter 2, the adoption of a networking strategy is also related to the internal development of environmental organizations. So before exploring how contextual conditions affect

Table 7.3 ENGOs and contextual conditions

	Political attitude	*Economic level*	*Local culture of utilizing personal networks*	*Degree of formalization*	*Extent of reliance on personal networks*
ASGP (Panjin)	Relatively closed	Low	Strong	Very low	Very strong
GH (Xiangfan)	Closed	Low	Strong	Low	Strong
GEV (Beijing)	Relatively open	High	Strong	Very low	Weak
Three cases (Shanghai)	Closed	High	Weak	High	Very weak

the utilization of personal networks, I first discuss the effect of formalization and professionalization of ENGOs on the use of personal networks.

What can be learned from the four cases is that the degree of formalization of the ENGOs has an inverse correlation with their reliance on personal networks. The higher the formality, the less likely it is that the ENGOs use informal mechanisms of personal networks to organize environmental activities, and *vice versa*. This means that personal networks should not be interpreted as an institutionalized mechanism in Chinese environmentalism, in an ahistorical way connected to Chinese conditions. Rather, it is more likely that the strong reliance on personal networks will partly disappear when these ENGOs develop into more formal, professional, and institutionalized organizations. But the ability to develop into more formalized environmental organizations remains at least partly dependent on contextual factors and cannot be completely seen as an endogenous process. Thus Chinese contextual specificities remain relevant in explaining the prevalence and continuity of personal networks within ENGOs.

Regarding the effect of the urban contextual conditions, first, the case studies provide some evidence that the attitudes and openness of political authorities do not always show a strict correlation with the adoption of personal networks. Especially the comparison of Beijing and Shanghai is telling in this regard, as the economic and size factors seem to be indiscriminate, while the political restrictions clearly differ between the two cities.[2] Despite the conventional wisdom that restrictive political attitudes and policies will stimulate and prioritize the development and utilization of personal networks, the case of Beijing seems to indicate that this strategy of personal networks is still used (although considerably less than in Panjin and Xiangfan) even when political authority is relatively open. And personal network strategy was not keenly adopted in the relatively closed and restrictive political environment of Shanghai.

When focusing on economic development in the four cities, the utilization of personal networks does appear to be correlated with economic development in urban China. In rich cities such as Beijing and Shanghai, personal networks are utilized to a lesser degree than in areas where economic growth is relatively slow and economic development levels are low, such as Panjin and Xiangfan. In poorer areas, the ENGO strategy tends to rely more heavily on personal networks in building consensus, influencing political systems, and establishing collective identities. One way of understanding this is through the prevalence of institutionalization in modern economic development processes. Economic development in modern market economies is generally achieved through established rules and institutions (North, 1990). In modern market economic development, relatively little space is available for using informal mechanisms in the economy, such as personal networks. This might very well spill over to other sectors. For instance, in Shanghai, environmentalists did not build *guanxi* networks with local political officials in order to mobilize campaigns protecting the local marshland. In poorer areas, such as in Xiangfan, local economic level reflects the institutional arrangements in the city, where informal mechanisms are still prevalent; whereas in highly developed and increasingly internationally integrated localities, the informal institutions are less prevalent in political and economic systems. Therefore environmentalists who intend to access political systems and build coordinated actions in Beijing and Shanghai rely less on informal institutional arrangements such as personal connections with local political officials.

Besides the political and economic conditions in each city, the cases show that the utilization of personal networks is related to cultural factors in each city. In cities where the culture of *guanxi* networks is prevalent, individuals are more inclined to make use of it. In Beijing, Xiangfan, and Panjin, where the subculture of using personal networks is strong, movement actors tend to be more inclined to utilize personal networks in disseminating collective identities. While in Shanghai, where local cultural heritage has diverted from traditional culture, this strategy is relatively less utilized in forming and disseminating shared ideas and values. Furthermore, this condition appears crucial for the adoption of *guanxi* networks in local green activism. In the case of Panjin, with no other alternatives, the leader of the ENGO had to deliberately develop *guanxi* connections in order to be able to mobilize coordinated actions. In the case of Xiangfan, although the ENGO has developed a higher level of formality than the previous case, and the environmentalists had also developed organizational networks, personal networks still proved irreplaceable in accessing government. This means that local culture is a prominent factor in affecting strategies that environmentalists adopt in mobilizing environmental reforms.

From the above analysis, we are unable to draw conclusions on the contextual factors that contribute to the use of personal networks. Our sample selection is far too small for a thorough factor analysis, and the contextual factors are also not completely independent. From the analysis of our case studies above, the best we can do is to formulate likely hypotheses. It seems likely that both cultural and economic factors are relevant contextual aspects in explaining variations in the reliance on personal networks, in addition to the level of formalization of ENGOs. Political restrictiveness seems less relevant with regard to variations in use of personal networks. Follow-up research is needed to distinguish how, and to what extent, these contextual aspects contribute to the use of personal networks (see below).

Answers to research questions

In this section, the research questions developed in Introduction are addressed.

- How do Chinese environmental NGOs use social networks in resource mobilization, influencing state policies and the realization of their goals?
- What is the specific development of Chinese ENGOs – *vis á vis* western ENGOs—in terms of their formalization and networks?
- Are the cultural features of Chinese society, or the early stage of ENGOs development, decisive for the utilization of personal networks as the main strategy of Chinese ENGOs?

In answering the question as to how Chinese environmental groups use social networks, the previous section analyzed in detail how personal networks fulfill various roles in environmental activism, whether or not to the same extent and the same way in all parts of the country and for all NGOs. Access to state authorities and the policy-making process, the formation of collective identities, and the mobilization of members towards collective actions draw significantly on personal networks. Organizational networks between movement organizations were (still) less prevalent and important (especially when compared with the dominancy of such networks among western environmental movement organizations), although certainly not absent in China.

The following section focuses in more detail on the similarities and differences found in the development of Chinese environmental movement organizations, compared with their Western counterparts, and thus answers the second research question.

Specific features of the Chinese environmental movement

As learned from the four cases studies, contemporary Chinese environmentalism distinguishes itself in two major ways from its counterpart in Western societies.

First, Chinese environmental organizations seem to follow different paths of development from their Western counterparts (Table 7.4). As analyzed in Chapter 2, this study distinguished four types of ENGO, with different levels of formality and types of strategy utilized. ENGOs of type I have a low level of institutionalization and formalization, and rely heavily on personal networks. The second type of ENGO has a similarly low level of formality and institutionalization, but operates largely through organizational networks and less

Table 7.4 Comparison between Chinese and Western ENGOs

	Early stage of NGO development	*Established NGO of more than 10 years' history*
China	I	I and II
Western OECD countries	I	IV

via personal networks. Type III has a high level of institutionalization and relies on personal networks in its activities; type IV shares the level of formality and institutionalization with type III, but works through organizational rather than personal networks. In contemporary, developed, Western society, type IV is dominant among influential national ENGOs. And the developmental path of ENGOs in Western industrialized societies predominantly goes from type I NGOs (low institutionalized organization, personal networks) to type IV (highly formalized, organizational networks).

The cases of ENGOs in this study demonstrate that Chinese ENGOs often have the same starting position when they are established: a low level of formality and a reliance on personal networks. But they have barely developed (yet) into formal organizations that rely predominantly on organizational networks (Table 7.5). In only one of the four cities did we see some indication of the establishment of type IV NGOs (Shanghai). This indicates that, at present, the final stage of the dynamic development of Chinese ENGOs largely differs from their Western counterparts. Only in the case of Shanghai was direct transformation from type I to type IV and from type III to type IV found. It is not coincidental that these organizations are strongly linked to, and supported by, international umbrella organizations that operate in Western OECD countries. In the case of Xiangfan and Beijing, the developmental path shows early signs of type III and type II, respectively, although it is too early to determine that these will become the dominant types for the next decade in these cities. In the case of the ASGP, the ENGO should still be considered a type I NGO. The question is whether these ENGOs will transform in the future into type IV NGOs. This relates again to our discussion about whether the specific (political, cultural, economic) conditions of China influence the level of formalization and use of personal networks; and also to the developmental path of China in the coming years (in terms of transition of its economic and political systems).

Second, the Chinese environmental movement in the case study cities adopts strategies that differ greatly from their Western counterparts. In Western European societies, professionalized ENGOs tend to utilize formalized political channels to influence policy-making; while less professionalized activists rely on direct actions to influence policy-making processes. In China, using informal networks (often with one central figure in the lead) appears to be the dominant prevailing strategy in accessing the political system and influencing policy processes. Personal networks are also crucial in organizing environmental campaigns and obtaining various resources, much more so than in Western NGOs.

Table 7.5 Organizational development from a dynamic view

	ASGP (northern China)	*GEV (Beijing)*	*GH (middle China)*	*WWF Shanghai (Shanghai)*	*SGSF (Shanghai)*	*GC (Shanghai)*
History	Very old (15 years)	Old (10 years)	Young (4 years)	Young (5 years)	Young (5 years)	Young (5 years)
Level of formality	Very low	Low	Medium	Low	High	High
ENGOs' development	I→I	I→I/II (start using organizational links)	I→I/III (signs of institutionalization)	I→II/IV (institutionalized; organizational network)	III→IV (personal networks of less prominent use than organizational networks)	III→IV (personal networks of less prominent use than organizational networks)

Personal networks: culture or development stage?

Social movement theories explain that strategies differ in the various stages of organizational development of NGOs. This postulate seems to be affirmed by this study on the Chinese environmental organizations: their low level of formality and institutionalization—characteristic for early stages in the development of social movement organizations—seems to go together with personal networks as the main informal strategy and coordination mechanism. But there is no linear time path to be seen in China. For instance, the GEV (established 10 years ago) and the ASGP (15 years ago), two of the oldest ENGOs in China, show no signs of developing into more formalized/institutionalized organizations. There are good reasons to believe that they will retain the same organizational structure for another five to ten years. If 20–30 years should therefore be regarded as an embryonic period for Chinese ENGOs, this means that the early development stage for Chinese ENGOs lasts much longer than for their Western counterparts. Part of the explanation for this relatively long embryonic stage for ENGOs in China can be found by looking at specific contextual factors, most notably the cultural, political, and economic specificities. There are, however, also signs that some Chinese ENGOs move much more quickly towards the path of formalization, professionalization, and institutionalization, often assisted by international donors and ENGOs.

The Chinese cultural context has a strong influence on the ample use of personal networks by ENGOs in mobilizing resources, organizing coordinated actions, and accessing the political system. As personal networks are also prevalent in other spheres, such as in the economy and the political system, it would be odd if they were absent in green activism. Thus green activism sees a strong reliance on informal mechanisms in China, which, to some extent, is independent of the development stage of ENGOs. We also found this in our case studies. Personal networks are of prime importance in cities that are least affected by strong international independencies and overseas influence, but that are dominated by the national Chinese culture.

So the best answer, based on the current empirical research, is that both the embryonic stage and the cultural context contribute to the strong reliance on personal networks—as opposed to organizational networks—in the contemporary Chinese environmental movement. Being mostly young organizations, Chinese ENGOs have the specific characteristics of all young or embryonic NGOs, among which are a stronger reliance on personal networks and a poorly professionalized, formalized, and institutionalized organization. But the fact that the embryonic phase sometimes lasts a very long time, that all sectors in Chinese society are permeated by these personal networks, and that in those locations where the specific Chinese culture is less compelling, organizational networks and professional organizations emerge, all show the importance of cultural factors in explaining personal networks in the Chinese environmental movement.

Agenda for future research

Our study, based on a few cities and ENGOs, has been largely exploratory in character. When the research design was developed at the start of the third millennium, Chinese ENGOs were not as widely spread or given as much room for maneuver as we see now. This new situation warrants a new research agenda. The follow-up to this research on Chinese environmentalism in a few locations should broaden the empirical base to a larger group of cities and ENGOs. By examining the role of personal networks in more localities in China, the initial findings of this study can be further validated and refined. There are four specific clusters of research questions that may comprise an inspiring future research agenda on Chinese environmental activism.

First, follow-up research is needed to distinguish the influence of Chinese cultural factors on the level of formality of NGOs in using personal networks. Such a study should be conducted in places where the influence of conventional Chinese culture is relatively weak, such as in Shanghai and other major cities along the southeast coast. With the ongoing proliferation of ENGOs throughout China, a quantitative research design is increasingly possible, preferably focusing on young ENGOs. If relatively weak evidence for the use of informal mechanisms is found among such ENGOs, it would point to the fact that cultural influences constitute an important factor in the prevalence of personal networks. If a majority of these young NGOs are found to rely strongly on personal networks, it can be concluded that the formality level plays a crucial role concerning this phenomenon.

Second, comparative research on the role of cultural and political factors in the use of personal networks by ENGOs would be interesting along two lines. Comparative research with other Asian countries, which share a more or less similar cultural background, but show different levels of economic and political development and different degrees of institutionalization/formalization of ENGOs, would be interesting. This could generate better insights on the specificities of cultural factors in the use of personal networks in green activism. The common cultural heritage with Taiwan, Korea, and to a lesser extent Japan would make them interesting comparative cases. In addition, comparative studies with economies in transition would be interesting, especially when there are similar levels of economic development. This could deliver better insight into the role of political conditions in the use of personal networks by ENGOs.

Third, this study provides anecdotal evidence that the adoption of strategies by Chinese NGOs is influenced by relations with international ENGOs. It would seem that the more connection and collaboration there is with international ENGOs, the more formalized and institutionalized the Chinese NGOs are. Specific research on the consequences of further integration of Chinese ENGOs into international/global civil society will provide insight on the specific relevance of the Chinese context in an increasingly globalizing civil society. Can we expect to see a further harmonization of Chinese ENGOs and Western NGOs, and to what extent?

Fourth, this study demonstrates that the Chinese environmental movement has had certain influences on the political system. Future research can investigate the specific influences that environmental movement organizations have on Chinese political authorities. Such studies are relevant in two ways: to obtain a better idea of the impact of green activism on the bureaucratic system; and on the ongoing power struggles within that bureaucratic–political system. Assessing the role that ENGOs play in the status and power of the Environmental Protection Bureaux in relation to other authorities in the Chinese political system is of key interest. In addition, the significance of environmental movement organizations on the first developments in Chinese democratization can also be explored by looking at developments in political participation via ENGOs. To what extent can the ENGOs be interpreted as new platforms of public participation, counterbalancing the political and economic interests of the powers that be? Or should (a part of) these ENGOs, with their still small constituency, and their personalized and partly closed networks, which sometimes comprise close relations with political economic elites, be interpreted as a component of these powers?

Research on Chinese green activism is only just starting. But given the size of the country, of its population, and of its (future) economic and political power, the future of Chinese environmentalism will be of relevance far beyond China. This makes this research agenda all the more pressing.

Appendix

Interviewees in this research

Staff of NGO

Interview with Mr Chen, project officer of Conservative International China, 22 January 2005.

Interview with Mr Guo, core staff of Grassroot Community, 20 May 2005.

Interview with Ms Ji, previous project officer of the 'environmental community' project in Grassroot Community, 12 March 2005.

Interview with Ms Jin, Director of Global Environmental Institute, former founder of Green Environmental Volunteers, 29 March 2005.

Interview with Ms Li, civil servant of middle rank from the Women's Association. She also served as Vice-President of the Association of Saunders's Gull Protection, 25 April 2004.

Interview with Mr Li, YN, retired head from Bureau of Education. He now serves in Green Han River's members' Committee, 11 September 2005.

Interview with Mr Li, ZH, Vice-President of Green Han River, 10 September 2005 and 13 September 2005.

Interview with Mr Liu, one of the main leaders of Grassroots Community, 1 January 2005.

Interview with Mr Liu, President of Association of Saunders's Gull Protection, 20 August 2005; 27 August 2005; 24 April 2004.

Interview with Mr Mao, previous program officer of Grassroot Community, 7 January 2005.

Interview with Ms Qian, Director of Center for Biodiversity and Indigenous Knowledge, 16 July 2004.

Interview with Mr Qiu, leader of Green Power, 2 January 2005.

Interview with Mr Qiu, President of Green Power, 8 February 2005.

Interview with Miss Shen, Director of Green Web, 4 April 2004.

Interview with Ms Song, Director of Shine Stone Community Participation Action, 15 June 2004.

Interview with Mr Sun, President of Green Cross, 21 December 2003.

Interview with Professor Tao, supervising teacher of the Urban Group in Shanghai Normal University, 9 January 2005 and 9 February 2005.

Interview with Ms Wang, Director of Green Environmental Volunteers, 12 November 2003 and 23 March 2004.
Interview with Mr Xu, program officer of Grassroot Community, 15 January 2005.
Interview with Mr Xue, Director of Friends of Nature, 2 January 2005.
Interview with Mr Yao, Director of Shanghai Wild Birds' Association, 20 May 2006.
Interview with Ms Ye, Vice-President of Green Han River, 6 September 2005.
Interview with Miss. Yu, project officer of Green Watershed, 17 July 2004.
Interview with Mr Yu, Director of Green Watershed, 17 July 2004.
Interview with Ms Yun, President of Green Han River, 19 December 2003 and 21 May 2004.
Interview with Ms Zhang, JZ, assistant project manager, Yunnan Environmental Development Programme (YEDP), 18 July 2004.
Interview with Ms Zhang, KJ, Director of Green Island, journalist of China Youth, 2 December 2004 and 23 March 2004.
Interviewed Ms Zhang, L, staff of Green Environmental Volunteers, 16 December 2003.
Interview with Mr Zhao, project officer of Green Environmental Volunteers, 24 March 2005.
Interview with Mr Zhu, Director of Grassroot Community, 1 January 2005.

Members and participants of ENGOs

Interview with Mr Fang, core activist from WWF Shanghai, 12 January 2005, 20 January 2005 and 16 May 2005.
Interview with Mr Li, member of Green Han River, 14 September 2005.
Interview with Mr Li, member of WWF Shanghai, 10 January 2005 and 29 January 2005.
Interview with Mr Li, TX. Qingfeng, core environmental activist of WWF Shanghai, 23 January 2005.
Interview with Mr Liu, former member of Green Student Forum of Shanghai, 8 January 005.
Interview with Mr Sun, member of the Council's meeting of Green Han River, 14 September 2005.
Interview with Mr Wu, member of Green Han River, 10 September 2005.
Interview with Ms Yan, participant of Green Union, Shanghai Maritime University, 20 January 2005.
Interview with Mr Yao, core member of WWF Shanghai, 13 January 2005, 22 January 2005, 4 April 2005 and 20 May 2006.
Interview with Mr Zhao, core member of Shanghai Wild Bird Protection Association, 12 January 2005 and 11 May 2005.

Government officials

Interview with Mr Leng, Vice-President of Panjin Bureau of Fishery. He also served as Secretary-General of Association of Saunder's Gull Protection, 26 April 2004.
Interview with Mr Li, officer in Natural Resource Protection Bureau, 27 August 2005.
Interview with Mr Li, Director of Natural Resource Protection Bureau, 27 August 2005.
Interview with Ms Qiu, Director of publicizing unit of Xiangfan City People's Political Consultative Conference, 9 September 2005.
Interview with Mr Wang, Vice-President of Xiangfan Environmental Protection Bureau, 15 September 2005.
Interview with Mr Yang, Director of Environmental Education in Xiangfan EPB, 15 September 2005.
Interview with Mr Zhu, Head of Panjin Environmental Protection Bureau, 25 February 2004.

Other actors

Interview with Ms Cai, journalist of Xiangfan Radio Station, 10 September 2005.
Interview with Mr Chen, journalist of *Xiangfan Daily*, 10 September 2005.
Interview with Miss Chen, *Xiangfan Evening News*, 13 September 2005
Interview with Mr Fu, Editor of *China Development Brief*, 16 June 2004 and 30 November 2004.
Interview with Mr Hu, environmental activist, 27 October 2003.
Interview with Mr Li, ZX, media reporter of Panjin TV Station, 24 August 2005.
Interview with Mr Liang, inhabitant of No. 710 Community, 19 January 2005
Interview with Mr Ren, manager of Dingxiang Group—a tourist company in Panjin, 29 August 2005.
Interview with Ms Shen, journalist from *Nanfang Weekly*, 20 January 2005.
Interview with Mr Wen, Director of Global Greengrants Fund, 5 October 2004 and 9 April 2006.
Interview with Mr Yu, Head of *Panjin Daily*, 28 May 2004.
Interview with Mr Zhang, manager of a local medical facilities company, 29 August 2005.
Interview with Mr Zhou, close friend of Ms Wang, 17 January 2005.

Notes

Introduction

1 Du, Jiangyan is an irrigation system that was constructed in the third century BC. This system still controls the water of the Minjiang River and distributes it to the fertile farmland of the Chengdu plains.
2 In March 2007, State Environmental Protection Agency changed its name to the Ministry of Environmental Protection of People's Republic of China.
3 'Green Ally—Government Cooperate with NGOs to Cope with Environmental Issues', *Xinhua News*, 21 June 2006. http://news.xinhuanet.com/politics/2006-01/21/content_4080350.htm
4 NGOs must have start-up funds of more than 100,000 yuan in the case of national organizations, while NGOs on the regional or local level must possess funds of more than 30,000 yuan. However, they are not allowed to raise funds until their application is accepted. In practice, this means it is possible to register an NGO if it is endowed with significant personal resources.
5 Interview with leaders of Green Han River, Mr Li and Ms Yun, 22 December 2003 and 5 September 2005, respectively.

1 Environmental governance and NGOs: an introduction to China

1 The new 'five-year' plan proposes that the goal will be reducing energy consumption per unit of GDP by 20 per cent of 2005 levels by 2010; and increasing renewable sources of energy to 5 per cent by 2010.
2 For instance, the State council promulgated 'Adopting Scientific Views by Strengthening Environmental Protection' in February 2006, strengthening the importance of improvement in environmental governance. The MEP and the Ministry of Supervision jointly promulgated 'Regulations on Violating Laws of Environmental Protection' in February 2006.
3 Sun, Xiuyan, 'To understand the first report on green GDP', *People's Daily*, 9 September 2006.
4 Available from news report on MEP website. www.mep.gov.cn
5 In China, so-called government-organized non-governmental organizations (GONGOs) form a special category of environmental groups with close relations to governmental institutions, but not part of the government.
6 Exception for the registration process exists, for instance, when social groups are considered as 'internal' organizations falling within an existing institution, organization or business enterprise: they do not have to register with the Ministry of Civil Affairs, and do not need an official sponsor any more. This can be important for many NGOs that have been able to establish themselves as programs associated with universities.

7 A different degree of autonomy exists for social organizations, under direction of their supervising organ. A first-grade social organization is allowed to have its own bank account; a second-grade social organization will not be allowed an independent bank account.
8 NGOs must have start-up funds of more than 100,000 yuan in the case of national organizations, while NGOs at the regional or local level must possess funds of more than 30,000 yuan. However, they are not allowed to raise funds until their application is accepted. In practice, this means it is possible to register NGOs if they have significant personal resources.
9 For example, one such NGO reports that in past years it consistently paid some nominal tax of a few hundred dollars annually. Another paid about 5 per cent of its income in taxes and 'sought the understanding' of the tax department.
10 Interview with Mr Chen, Project Officer of Conservative International China, 22 January 2005.
11 Because of the limitation of this survey, the number of unregistered ENGOs may not be exact.
12 This survey was taken by a GONGO, ACEF, which may have had difficulties trying to reach unregistered grassroots ENGOs that do not have legal status.
13 Interview with Mr Zhou, scholar from Renmin University, 17 January 2004.
14 Yuankai, Tang, 'International NGOs play important role in China', *Beijing Review*, No. 22, 2006, 13–16.
15 Interview with Mr Chen, Project Officer of Conservative International China, 22 January 2005.
Interview with Mr Wang, Director of Oxfam, 22 January 2005.
16 Social security in China normally includes medical insurance, unemployment insurance, old-age insurance, housing insurance, etc.
17 Interview with Ms Shen, Director of Green web, a grassroots NGO in Beijing, 7 March 2004.
18 This relationship between affluence and environmental awareness might be very different when other environmental indicators are used, such as clean drinking water, erosion or water-borne diseases.
19 Interview with Ms Xu, Vice-Director of Center for Legal Assistance to Pollution Victims in China, 15 November 2003.
20 In a document promulgated by the State Council in December 2005, 'State Council's Decision on Strengthening Environmental Protection by Scientific Views', social organizations are encouraged to participate in supervising environmental projects.
21 Bin, Ouyang, 'Difficulties for NGOs and Civil Societies in China', 7 September 2005. www.threegorgesprobe.org/gb/index.cfm?DSP=content&ContentID=13675
22 Song, Zonghe, 'GONGO and grassroot NGOs—cooperation just begins', *China Philanthropy Times*, 11 June 2006.
23 Available from China Development Brief website, section on international NGOs. www.chinadevelopmentbrief.com
24 *Ibid.*
25 Interview with Mr Wen, Director of Global Greengrants Fund, 5 October 2004.
26 Interview with Ms Ji, Pre-program Officer for Environment Protection in Grassroots Community, Shanghai, 10 January 2005.
27 21st Report on Internet development in China [*Di ershiyi ci zhongguo hulianwang fazhan tongji zhuangkuang baogao*], conducted in 2007. www.cnnic.net.cn/index/0E/00/11/index.htm

2 Analyzing Chinese environmental movement networks

1 The five principles are the main objects for measuring networks in relational analysis. Other principles are also mentioned, depending on the purpose of network

analysis. Several methodologies and methods have been developed to quantify and measure these five principles (Wasserman and Faust, 1994).

2 *Ren* requires the norms of behavior that favors people with whom one has a close relationship; *yi*, or righteousness, demands respect to those for whom respect is required by the relationship; and acting according to previously established rites or social norms is called *li*.

3 Environmental activism in Panjin, Liaoning Province

1 Panshan and Dawa as counties, and Shuangtai zi and Xinglong as districts.
2 Panjin Government's report of 2004. www.liaohe.net.cn
3 *Ibid.*
4 Available from the Panjin City Government's website. www.panjin.gov.cn
5 Available from *China Statistical Year Book 2003* and *Panjin Statistical Year Book 2003*.
6 Interview with Mr Zhu, head of local Environmental Protection Bureau, 25 February 2004.
7 Hotline calls '12369' were established in 2001. By 2003, hotlines had been installed in 308 cities for reporting environmental emergencies across the country; 290,000 cases were reported in these two years. The average calls in each city (or region) reached 471. From news section of the Ministry of Environmental Protection Agency (MEP). www.mep.gov.cn.
8 Available from www.birdlife.org/datazone/species/index.html.
9 *Ibid.*
10 Available from www.ramsar.org.
11 Internal report of Mr Li, Director of NRPB, August 2005.
12 An international convention that was adopted in Ramsar, February 1971, now signed by over 110 countries, legally binding countries to promote the wise use and conservation of wetlands. www.ramsar.org
13 Report on the 'Construction of Panjin City Ecological Monitoring Station', conducted by Panjin Meteorological Office, 27 October 2005.
14 Sun, Yanfei and Li, Haipeng, 'How could nature reservation solve its management dilemma?' *Nanfang Weekly*, 4 March 2004.
15 They are two farms of bulrushes in the reservation park, both independent administrative areas in Panjin city.
16 Sun, Yanfei and Li, Haipeng, *op. cit.*
17 Collective membership is one way to mobilize young children and students to participate collectively. It is based on organizational linkages between ASGP and the schools.
18 Interview with Mr Liu, President of ASGP, 20 August 2005.
19 Interview with Mr Liu, member of ASGP, 27 August 2005.
20 Interview with Mr Leng, member of ASGP, 24 May 2004.
21 Interview with Mr Liu, Director of ASGP, 20 August 2005.
22 Interview with Mr Yu, former member of ASGP, 2 February 2005.
23 *Ibid.*
24 Interview with Mr Liu, President of ASGP, 24 April 2004.
25 Interview with Mr Yu, Head of *Panjin Daily*, 28 May 2004.
26 Interview with Mr Liu, President of ASGP, 24 April 2004.
27 Interview with Ms Li, civil servant of middle rank from the Women's Association (she also served as Vice-President of ASGP), 25 April 2004. Interview with Mr Leng, Vice-President of Panjin Bureau of Fishery, who also served as General Secretary of ASGP, 26 April 2004.
28 Interview with Mr Ren, Manager of Dingxiang Group—a tourist company in Panjin, 29 August 2005. Interview with Mr Zhang, manager of a local medical facilities company, 29 August 2005.

29 *Mianzi* represents social perceptions of a person's prestige. For a person to maintain face is important in Chinese social relations because face translates into power and influence. A loss of *mianzi* would probably result in a loss of authority.
30 Interview with Mr Wu, retired official of NRPB, 27 May 2004.
31 Interview with Mr Li, media reporter of Panjin TV Station, 24 August 2005.
32 Interview with Mr Wu, retired official from the NRPB, 27 May 2004.
33 This position is in the vertical line of the party system in Chinese bureaucracy. It ranks at the same level as a mayor of a city. However, because of the dominance of CCP, this position has more concrete power in reality.
34 Xia, Hua, 'Couple loves bird and cares for Saunders's Gull', *Panjin Evening News*, 10 July 2002; Huang, Shisheng, 'Such citizens are needed for constructing ecological city', *Panjin Evening News*, 15 March 2004.
35 Gao, Shuhua, 'Sui, Xiaoqing being interviewed by central TV station', *Panjin Daily*, 21 November 2003.
36 The main articles on Saunders's Gull include: Liu, Detian, 'Talks about "The bird as the symbol of Panjin"', *Panjin Daily*, 24 October 1995; 10 October 1995; 26 September 1995; 12 September 1995; 29 August 1995; 4 July 1995; 'Candidate birds as the bird as city symbol', *Panjin Daily*, 23 August 1995; 'Which can be the bird as the symbol of the city?' *Panjin Daily*, 20 July 1995; 'Where are you—the bird to be the symbol of Panjin?' *Panjin Daily*, 26 May 1995; Editor, 'Collective participation in the selection of the bird as the symbol of the city', *Panjin Daily*, 10 July 1995.
37 Another kind of bird, the red-crowned crane, was also elected to be bird of the city.
38 Xia, Hua, 'Ceremony for Saunders's Gull', *Panjin Daily*, 2 July 2002; Zheng, Baiqing, 'Opening ceremony of the Panjin association of Saunders's Gull Protection', *Panjin Daily*, 21 June 2003.
39 Yu, Sen, 'Research Staff Investigating the Association of Saunders's Gull Protection', *Panjin Daily*, 26 May 2004.
40 Wang, Hanlin, 'God of protecting birds—story of Liu Detian and Saunders's Gull protection', *China Environment News*, 27 August 2002; Zhang, He, 'Liu, Detian's love on birds'; *People's Daily*, 31 July 2003; Li, Yuyang, 'Liu, Detian: to save Saunders's Gull', *Economy Daily*, 8 July 2005; Xia, Jun, 'Liu, Detian—the 2001th Saunders's Gull'; *People's Daily*, 7 August 2005; Jiang, Min and Wang, Wei, 'Liu, Detian spent 15 years to protect birds', Xinhua News Agency (Shenyang), 28 June 2005; Xia, Hua, 'Liu, Detian makes Panjin journalist known to the nation', *Panjin Daily*, 7 June 2002.
41 The yearly counting is organized by the NRPB; data obtained from Xiao, Chong, 'Attention to the Wetland and Black-beak Gulls', *Panjin Evening News*, 11 July 2003.
42 Yu, Sen, 'The death of 2,000 Saunders's Gulls', *Panjin Daily*, 3June 2003.
43 Liu, Detian, 'Crying small South river', *Panjin Daily*, 12 July 2003; Yu, Sen, 'Experts from three countries gather in Panjin to investigate Saunders's Gulls' living', *Panjin Daily*, 11 July 2003; Xiao, Chong, 'Attention to the wetland and black-beak gulls', *Panjin Evening News*, 11 July 2003; Xia, Hua, 'The loss of Saunders's Gulls alerted journalist of Renmin Daily', *Panjin Evening News*, 11 July 2003; Yu, Sen, 'This year's Saunders's Gull is 2,000 less than last year', *Panjin Evening News*, 2 July 2003; Yu, Sen, 'Hundreds nests being flushed along Small South River Bank', *Panjin Evening News*, 25 June 2003; Yu, Sen, 'The silence of the Japanese expert', *Panjin Daily*, 25 June 2003; Liu, Detian, 'Attention to the homeland of Black-beak Gulls', *Panjin Daily*, 13 June 2003; Liu, Detian, 'Saunders's Gull died in tragedy around Small South river', *Panjin Daily*, 3 June 2003; Liu, Detian, 'Xiaonan River to be protected as Saunders's Gulls' homeland', *Panjin Daily*, 2 April 2003.
44 Zhang, He, 'Liu, Detian's love on birds', *People's Daily*, 31 July 2003.
45 Interview with Mr Li, officer in NRPB, 27 August 2005.
46 *Ibid.*

4 Environmental activism in Beijing

1 Available from Beijing Statistical Information Net. www.bjstats.gov.cn
2 'GDP per person reached 5,000 dollars that represents the income level of medium developed countries', *Beijing Morning Post*, 23 January 2006.
3 Information gained from Beijing Government website on special ecological planning for the Olympic Games. www.beijing-2008.org/30/90/article211989030.shtml
4 Annual Environmental Indicators in Beijing, available from the website of Beijing Environmental Protection Bureau. www.bjepb.gov.cn
5 Information gained from Beijing Government's website on special ecological planning for the Olympic Games. www.bjoe.gov.cn/ayybj/aygh/zxgh/200508/t44480.htm
6 Annual Environmental Indicators in Beijing, *op. cit.*
7 Information gained from Beijing Government's website on special ecological planning for the Olympic Games, *op. cit.*
8 Hildebrandt, T. and. Turner, J. L. 2003, 'A roundtable discussion with MEP minister Xie Zhenhua', 9 December 2003, available from website of Woodrow Wilson International Center for Scholars. http://www.wilsoncenter.org/index.cfm?topic_id=1413&fuseaction=topics.event_summary&event_id=53003
9 Interview with Mr Fu, Editor of *China Development Brief*, 30 November 2004.
10 News report, 'Behind the Storm Led by SEPA', 21 January 2005. www.cfej.net
11 Learned from website of MEP. www.mep.gov.cn
12 Interview with Mr Xue, Director of Friends of Nature, 2 January 2005.
13 Interview with Mr Sun, former project officer in Global Village Beijing, 21 December 2003. Interview with Ms Zhang, staff of GEV, 16 December 2003. Interview with Mr Xue, Director of Friends of Nature, 2 January 2005. Information on GEV is learned from Mr Zhao, project officer of GEV, 24 March 2005.
14 Interview with Ms Jin, Director of Global Environmental Institute, 29 March 2005.
15 Interview with Mr Wen, Director of Global Greengrants Fund, 9 April 2006. Interview with Mr Sun, former project officer in Global Village Beijing, 21 December 2003.
16 Available from the website of Green Island. www.cydgn.org
17 Interview with Ms Song, Director of Shine Stone Community Participation Action, 15 June 2004.
18 Interview with Ms Zhang, journalist of China Youth and Director of Green Island, 2 December 2004. Interview with Mr Fu, Editor of *China Development Brief*, 16 June 2004.
19 Interview with Ms Song, Director of Shine Stone Community Participation Action, 15 June 2004.
20 Interview with Ms Zhang, Director of Green Island, 2 December 2004. Interview with Mr Wen, Director of Global Greengrants Fund, 5 October 2004.
21 Interview with Mr Wen, Director of Global Greengrants Fund, 9 April 2006. Interview with Miss Shen, Director of Green Web, 4 April 2004.
22 Interview with Mr Hu, environmental activist, 27 October 2003. Interview with Mr Wen, Director of Global Greengrants Fund, 5 October 2004.
23 Both are small environmental groups from the USA.
24 Interview with Ms Jin, former founder of Green Volunteers Beijing, 29 March 2005.
25 Journalist Saloon, a program organized by GEV, formed such an occasion where news was released, ideas exchanged and campaign mobilization took place.
26 Interview with Mr Zhao, project officer of GEV, 24 March 2005.
27 The reasons for Ms Jin's leaving were various. She was offered an opportunity to establish another environmental NGO that focused on utilizing expertise in environmental reforms. Besides, her ideas differed from Ms Wang in the development of the GEV, which also contributed to her leaving.
28 Interview with Mr Zhou, close friend of Ms Wang, 17 January 2005.

29 She is a famous writer who participated in 1989's Student Movement, and is active in posing controversies toward the Chinese Government.
30 The data are gained through Ms Wang's rough estimation in January 2005. Her personal observations on the various activities of GEV during the Nu River dam campaign are adopted, and the core members of GEV (around 150) who joined its activities more than three times since 1997 are counted.
31 Available from *2005 Yearbook of Chinese Cities.*
32 The 'journalist salon' consists of regular meetings to help educate journalists on how to improve their coverage of environmental issues; bird watching is to observe birds through trips and expeditions.
33 Interview with Ms Zhang, journalist of China Youth and Director of Green Island, 2 December 2004.
34 See the discussion in Friends of Nature's newsletter (February 2003), in which the spirit is explained by Li, Lailai—member of Friends of Nature, also the Director of IED—in her letter responding to a young man's remarks on environmentalists.
35 Some of the details of the debate and disagreements between the authorities on the Nu River projects can be found in 'Opinions on development over issue of Nu River', *China Youth Daily*, 19 April 2004.
36 Compared with the investment of US$1.15 billion it received from the central government over the past 50 years, this opportunity was believed to be a turning point for the Nujiang area. It was estimated that the construction of the dams would bring around US$1.25 billion to Nujiang local government per year, and a total investment of US$105.9 billion. The construction means a transformation in the local economy, with tremendous growth of local income and employment. The energy produced by the dam is estimated to be 1.22 times the electricity produced by the famous Three Gorges dam in the Yellow River.
37 Yardley, Jim, 'China's premier orders halt to a dam project threatening a "lost eden"', *New York Times*, 2 February 2004.
38 Interview with Miss Yu, project officer of Green Watershed (Kunming), 17 July 2004.
39 Interview with Ms Zhang, Assistant Project Manager, Yunnan Environmental Development Programme (YEDP), funded by the UK Government's Department for International Development, 18 July 2004. Interview with Miss Yu, project officer of Green Watershed (Kunming), 17 July 2004. Interview with Mr Yu, Director of Green Watershed, 17 July 2004.
40 Interview with Mr Yu, Director of Green Watershed, 17 July 2004.
41 Interview with Ms Qian, Director of Cibik, 16 July 2004.
42 Interview with Mr Yu, Director of Green Watershed, 17 July 2004.
43 Interview with Ms Zhang, Assistant Project Manager, YEDP, 18 July 2004.
44 The National Investigation Institute, an institution closely linked to NDRC, and Kunming Investigation Institute, a local government institution, completed the environmental assessment. It was later regarded by MEP and environmental NGOs as neither neutral nor reliable.
45 As remarked by Mr Pan, the head of MEP, NGOs are allies of MEP in environmental protection matters.
46 The Chinese mass media have a modeling effect. The local media receive news from several main news agencies. There would be hardly any consensus if the news was reported first by national news agencies.
47 Interview with Ms Zhang, journalist of China Youth and Director of Green Island, 23 March 2004.
48 Zhao, Yongxin, 'Keep the river of ecological significance for future generations', *People's Daily*, 11 November 2003. Zhang, Kejia, 'Dam building on Nu River is planned; construction is opposed', *China Youth Daily*, 19 August 2003.
49 Ma, Hongtao, 'Conflicts in dam building on Nu River', on *Half an Hour on Economy*—TV program by National TV Station, 28 December 2003.

50 These were reported in articles in this period. Wang, Yongchen, 'Please keep the last river that has ecological significance', *Nanfang Daily*, 20 August 2003; Wang, Yongchen, 'Please keep the last river that has ecological significance', People's Consultancy News, 26 August 2003; Ding, Pin, 'Ecology in Nu River should be kept', *Chinese Environmental News*, 5 September 2003; Zhou, Yuanchun, 'Can Nu River skip the fate to be exploited?' *Science Newspaper*, 14 September 2003; Wang, Yongchen, 'Please keep the last river that has ecological significance', *Xinmin Newspaper*, 11 July 2003; News report, 'Nu River—the last river not being exploited is in danger', http://phtv.ifeng.com; Zhang, Hui'e, 'Dam will be built on Nu River too', *Nanfang Daily*, 25 November 2003.

51 Jin, Xiaodong and Xu, Hualin, 'Experts talk about gains and loss on dam building on Nu River', Radio Programme on National People's Radio, 4 September 2003; Ding, Pin, 'Scholars advocating protecting Nu River', *Chinese Environmental News*, 9 September 2003; Zhang, Kejia, 'Scholars collectively oppose dam building on Nu River: the last river of ecological significance', *China Youth Daily*, 9 September 2003.

52 Relevant reports, see Tang, Jianguang, 'Force that delayed construction on Nu River.' *China News Weekly*, 20 May 2004. Liu, Chang, 'Visit Nu River', *International Tribune*, 19 April 2004.

53 Interview with Ms Wang, Director of GEV, 23 March 2004.

5 Environmental activism in Xiangfan

1 The Xiangfan administration is composed of Nanzhang County, Gucheng County, Zaoyang County, Yicheng City, Laohekou City, Xiangyang District, Fancheng District, Xiangfan Yuliangzhou Tourism Economic Development Zone, Xiangfan Automatic Industry Economic & Technological Development Zone.

2 Statistics on Xiangfan's Economy and Social Development, 2004. Available from website of Hubei Provincial Bureau of Statistics, www.stats-hb.gov.cn

3 Hubei Development and Reform Commission's report: 'Industrial Development Plan in 10th 'Five-year Plan in Hubei Province'. www.hbjw.gov.cn

4 Available from Hubei Provincial Bureau of Statistics. www.stats-hb.gov.cn/tjj/

5 Available from the website of Hubei Provincial Bureau of Statistics, 'Achievements and Problems in the Industrial Structure Adjustment in Hubei Province'. www.stats.gov.cn/was40/reldetail.jsp?docid=402268992

6 This is a large-scale policy that is initiated by central government. The plan is to divert 13.4 billion cubic meters (12 trillion gallons) of water a year from the Yangtze River in China's wetter south to arid areas in the north. The water shortage of Beijing and Tianjin is expected to be lessened.

7 Interview with Mr Chen, journalist of *Xiangfan Daily*, 10 September 2005.

8 Collective membership is one way to recruit participants collectively. It is based on organizational links between GH and local government organizations, schools and business institutions.

9 The GH is financially supported by government agencies for their cooperating activities.

10 The data is retrieved from the members' registration book.

11 Interview with Mr Wu, member of GH, 10 September 2005.

12 Members' articles published on the website of GH. www.greenhj.org/manage/xinsheng/newshtm/2005929111116.htm

13 For example, members are required to work well in their working places.

14 Interview with Mr Li, Vice-President of GH, 10 September 2005.

15 *Ibid.*

16 Interview with Mr Sun, Member of the Council's meeting, 14 September 2005.

17 Available from GH's website. www.greenhj.org

18 Interview with Mr Li, Vice-President of GH, 10 September 2005.

19 Ms Yun is an active actor in local politics. She participated in the mayoral election in 2001. The mayor's term of office lasts five years. She was recommended as member of the XCPPCC since 1984 for four rounds; for two of them she served as a member in the permanent committee of the XCPPCC. This committee functions as an alternative organ of XCPPCC when its meeting stops. It meets every three months to discuss major policies and social issues.

20 In China, *danwei* refers generally to a special mechanism under the centrally planned economy. This phenomenon is becoming less apparent nowadays, since the market economy was developed. Yet it is still prevalent in Xiangfan.

21 Interview with Ms Yun, leader of GH, 21 May 2004.

22 Ms Ye still served as a member of the XCPPCC in 2005.

23 Interview with Mr Li Yn, retired head of the Bureau of Education (who now serves in GH's Members' Committee), 11 September 2005.

24 *Ibid.*

25 Interview with Ms Ye, Vice-President of GH, 6 September 2005.

26 Interview with Ms Yun, President of GH, 19 December 2003.

27 Usually, finding supervising institutions poses a major difficulty for ENGOs.

28 Three months is a relatively short period. In many cases, six months is needed. Social groups have to fulfill certain requirements. And finally, the agency of Civil Affairs determines if a social group is to be legalized.

29 Interview with Mr Li, Vice-President of GH, 13 September 2005.

30 CPPCC consists of both Party members and non-Party members. The members are chosen by the CPC, but are selected from a somewhat broader range of people than are normally chosen for a government office. The XCCPPC is organized above county level in all administrative levels. This organ meets at least once a year at the local level. The members' term of office lasts five years.

31 Internal publication: Peng Zhaoyin, Head of the City EPB, 'Direct Our Work with Scientific Views in Environmental Protection', 30 May 2005.

32 Its programs are broadcast twice a week on a local TV station.

33 Interview with Ms Qiu, Director of the publicizing unit of XCPPCC, 9 September 2005.

34 Interview with Mr Wang, Vice-President of Xiangfan EPB, 15 September 2005.

35 Peng Zhaoyin, Head of the City EPB, 'Direct our work with scientific views in environmental protection', 30 May 2005.

36 Yang Gengyun and Xiao Fang, 'Trial of political responsibilities in environmental governance', *Hubei Daily*, 22 June 2005.

37 Interview with Mr Wang, Vice-President of Xiangfan EPB, 15 September 2005.

38 Peng, Zhaoyin, head of the City EPB. 'Direct Our Work with Scientific Views in Environmental Protection', 30 May 2005.

39 Yang Gengyun and Xiao Fang, 'Trial of political responsibilities in environmental governance', *Hubei Daily*, 22 June 2005.

40 Peng, Zhaoyin, 'Direct our work with scientific views in environmental protection', 30 May 2005.

41 Interview with Miss Cai, journalist of Xiangfan Radio Station, 10 September 2005.

42 Interview with Mr Chen, *Xiangfan Evening News*, 13 September 2005.

43 Organizational links of GH are made up of the following ENGOs: Friends of Nature (in Beijing), Global Village Beijing, Global Environmental Volunteers (in Beijing), Institute of Environmental Development (in Beijing), Beijing Earthview Environment Education and Research, Chongqing Green Volunteer Union, Green Stone (in Nanjing), Friends of Green Environment of Jiangsu (in Nanjing), Tianjin Friends of Green, Green Beijing, China Development Brief (in Beijing), Yunnan Econetwork, Greenweb (in Beijing), Community Action (in Beijing), Friends of Earth Hong Kong, Green River (Xichuan), Shanxi Mother Environment Volunteer Association (in Xi'an).

44 Friends of Nature conducted its educational program—Green Hope Action has been conducted for three years in Xiangfan.
45 Three volunteers from GH were sent to Beijing in 2004, to participate in a program called 'Training for Mobile Environmental Education'. In such cases, Friends of Nature usually had to provide financial support to its participants.
46 Annual Report of GH, from website of GH. www.greenhj.org
47 In 2004, four NGOs (Friends of Nature, Global Village Beijing, Institute of Environmental Development, and China Association for NGO Cooperation) built a link to GH's homepage on their websites.
48 Li Zhihe, 'Ten things not known by Xiangfan people about South–North river diversion', *Newsletter of Friends of Nature*, 4, 10–12.
49 Interview with Mr Li, member of GH, 14 September 2005. Interview with Ms Yun, President of GH, 21 May 2004.
50 *Ibid.*
51 Available from GH's working report 2004, from website of GH. www.greenhj.org
52 *Ibid.*
53 Interview with Mr Wang, Vice-President of Xiangfan EPB, 15 September 2005.
54 Interview with Mr Yang, Director of Environmental Education in Xiangfan EPB, 15 September 2005.
55 The target audience is policy-makers and political leaders in all levels of government. The purpose was to educate civil servants in environmental protection. A greater part of the training was dedicated to introducing the major legal documents about environmental protection. Yet the lectures were organized without fixed dates. Interview with Mr Yang, Director of Environmental Education in Xiangfan EPB, 15 September 2005.
56 *Ibid.*
57 Interview with Ms Ye, Vice-President of GH, 6 September 2005.
58 The document promulgated is 'Requirement to Conduct Environmental Education in Middle Schools, Primary Schools and Kindergartens', jointly issued by Xiangfan Education Bureau and Xiangfan EPB, 15 October 2003.
59 Interview with Ms Yun, President of GH, 18 December 2003.
60 Available from GH's website. www.greenhj.org
61 The five communities are Wangfukou, Tanxicun, Shuixingtai, and Gaozhuang Residential Community. They are all located in urban areas.
62 Interview with Mr Sun, President of GCR, 21 December 2003.
63 An evaluation on this project was conducted by scholars and master's students from the Department of Sociology of Beijing Normal University. www.npo.org.cn
64 Yang Gengyun and Xiaofang, 'Trial of political responsibilities in environmental governance', *Hubei Daily*, 22 June 2005.
65 Xinhua Agency, 'Paper factory discharge wastewater in upper river, lower river becomes village of cancer', 20 August 2005. http://news.xinhuanet.com/mrdx/2005-08/20/content_3380304.htm
66 *Ibid.*
67 Interview with Ms Qiu, Director of publicizing unit, XCPPCC, 9 September 2005. Interview with Ms Yun, Director of GH, 21 May 2005.
68 In both April and December 2004, the Environmental Protection Team disseminated environmental education to villagers living along Tangbai River.
69 More than 10 media republished these articles. For example, *Changjiang Daily*, *Supervision Daily*, etc.
70 The article written by Duan Zekun is 'I Wish to Swim with Fishes in the Tangbai River'. Another introductory article was posted along with it: Chen Chunlin, 'Discharge in Upper River, Fishes Dead out in Lower Part', *Xiangfan Daily*, 29 April 2005.
71 The two newspapers reported the news on 19 July 2005 and 29 August 2005, respectively.

72 For example, 'Traveling to Han River in the Eve of Lunar year', an article written by Wang Yongchen was posted on the homepage of one ENGO—Green SOS. http://greensos.eedu.org.cn/info/news/home/200601/1019.html; www.nujiang.ngo.cn/Dynamics/116/view 1-07-2005 The website www.nujiang.ngo.cn was one of the important sources for sharing information in the campaign of anti-dam construction campaign. It attracts environmental activists and ENGOs, through which successful campaigns were built.
73 For example, 'Walks by the shore of Han River at the eve of the new year', an article written by Wang, Yongchen was posted on www.nujiang.ngo.cn, one of the important sources for sharing information in the campaign of anti-dam construction campaign. It attracts environmental activists and ENGOs, through which successful campaigns were built.
74 Referred to GH's working report of 2005, www.greenhj.org. More than 20 members joined the program. This was a 25-minute program, which was broadcast as a repeat eight times in 2005.
75 Yang Huayun, 'Ecological crisis led by hundreds death in village', *Oriental Daily*, 21 October 2005.
76 Yang Hong and Zhang Wenhua, '11-year-old boy impressed Xiangfan City—stories about Duan, Zekun, student of Migong primary school, Xiangfan', *Xiangfan Daily*, 27 January 2006.
77 To complement the government's responsibilities was one of the considerations in the GH's policy-making process. It very often consulted with the authorities concerned before it began initiating its activities.

6 Growing environmental activism in Shanghai

1 'Ranking of Chinese cities.' http://economy.enorth.com.cn/system/2006/01/10/001207768.shtml. Shanghai Bureau of Statistics. www.stats-sh.gov.cn/2004shtj/tjgb/tjgb2005.htm
2 During the first eight years after 1949, the income of the city government was submitted directly to the central government. From 1958 to 1987, the central government collected a fixed amount of revenue, which provided more incentives for local government to boost economic development.
3 Lin Yongting, 'No maximum should exist for resource distribution', *Shanghai State Asset*, 3, 21-25.
4 It is subordinated to the Bureau of Civil Affairs, as an associate bureau. As the primary government agency regulating NGOs, SBCO is sufficiently equipped with facilities and financial input. Established 1999, 10 million RMB was invested in the construction of this agency.
5 Available from Shanghai Environmental Protection Bureau's website, www.sepb.gov.cn; and Shanghai Civil Affairs Bureau, www.shmzj.gov.cn.
6 Dong, Qiang, 'Mayor Chen: development of civil society organizations is unavoidable trend of modern society', *Liberation Daily*, 25 August 2005.
7 Interview with Mr Fu, Editor of *China Development Brief*, 6 June 2004.
8 A thousand web cafés were pulled down by local government officials in Shanghai, which were accused to have corrupted local youth by providing access to 'primitive' (pornographic and/or subversive) online content. Lin Neumann, 'The Great Firewall', *Press Freedom Reports*, 3 June 2000, www.cpj.org/Briefings/2001/China_jan01/China_jan01.html. As part of a campaign combating 'harmful internet contents' launched on 1 May 2001, the Shanghai police had closed nearly 200 internet bars in the city. The Trade and Industry Bureau had judged 'socially very effective' the closure of internet bars not holding licenses. According to the official media, 200,000 internet bars were listed at the end of April in China, and 17,000 had been closed for 'illegal activities'. News report from Reporters

Without Borders, 'Closure of more than two thousand internet bars', www.rsf.org/article.php3?id_article=2624

9 Howard W. French, 'As Chinese students go online, little sister is watching', *The New York Times*, 9 May 2006.

10 According to the Shanghai Year Book of 2002, 28.9 per cent of Shanghai's population used the internet in the year 2000; 53.8 per cent of the population indicated they planned to establish internet facilities at home.

11 Some environmental activists report that they access news mainly through the internet, rather than from newspapers and conventional mass media.

12 Wang Dandan, 'Shanghai internet users amount the most across China', 21 January 2005. http://media.people.com.cn/GB/40606/3135370.html

13 On average, 17 per cent of the urban population utilizes the internet in China. Data from China Internet Network Information Center. www.cnnic.net.cn/html/Dir/2006/01/17/3508.htm

14 Interview with Mr Xu, program officer of GC, 15 January 2005.

15 Lin Wei, 'From online till offline—new type of association attracts Shanghai youth', *China Youth Daily*, 12 December 2005.

16 Available from *Shanghai Environmental Bulletin* (*Shanghai Huanjing Gongbao*) on Shanghai Environmental Protection Bureau's website. www.sepb.gov.cn

17 *Ibid.*

18 The goal of 'ecological city' refers primarily to the sustainable development of a city. The criteria to judge if a city is ecological include several aspects, including a rational plan for the city's development, pleasant urban ecological environment, sanitary city environment, less pollution, environmental protection, traffic system, etc. Available from Shanghai Environmental Protection Bureau's website. www.sepb.gov.cn

19 For instance, in order to clean the Huangpu River, a special office—the Office for the Protection of Shanghai Huangpu River Source—was established, under the Shanghai EPB, with special responsibility for the management and enforcement of pollution control in the upper reaches of the Huangpu River.

20 Monthly journal reported by MEP's Center of Environmental Education. www.chinaeol.net/xjwl/hjxcjb.asp

21 The Agencies include the Shanghai Municipal Development and Reform Commission, the Shanghai Municipal Economic Commission, the Shanghai Municipal Science and Technology Commission, the Shanghai Municipal Bureau of Public Security, the Shanghai Municipal Finance Bureau, the Shanghai Municipal Commission of Construction and Administration, the Shanghai Municipal Public Health Bureau, the Shanghai Municipal Water Affairs Bureau, the Shanghai Municipal City Planning Administration, the Shanghai Municipal Urban Communications Administration Bureau, the Shanghai Municipal Government Offices Administration, the Shanghai Municipal Public Works Administration, the Shanghai Municipal City Appearance and Environmental Sanitation Administration, the Shanghai Municipal Agricultural Commission, the Shanghai Harbor Administration, the Shanghai Bureau of Justice, the Shanghai Municipal Transportation Bureau, the Shanghai Municipal Oceanic Bureau, the Shanghai Municipal Maritime Bureau, the Shanghai Municipal Commission of City Greening Administration, the Shanghai Municipal Administrative Office of Suzhou River, the Shanghai Municipal Administration of Wusong Industrial Zone, the Shanghai Municipal Administration of Riverbanks, the Shanghai Commission of Industrial Districts, the Bao Steel Corporate (a state-owned company), and the Construction office of the New Harbor.

22 Chinese bureaucracy is structured in vertical lines and horizontal sections. Governmental units in line have no authority in directing units in horizontal sections. This is also the case in environmental governance. The Environmental Protection

Bureau is authorized to give directions to EPBs in district level in a vertical line. But it has no authority over other governmental units in district level in horizontal sections.

23 Xu Zhenglong, Liu Youbin and Zhao Guanliang, 'Reports on Shanghai Municipal's Environmental Protection', *China Environmental News*, 2 November 1999.

24 'Shanghai's environmental protection and ecological construction brief', available from Shanghai Environmental Protection Bureau's website. www.sepb.gov.cn

25 *Ibid.*

26 Available from Shanghai Environmental Protection Bureau's website, www.sepb.gov.cn, and *Shanghai Environmental Bulletin* (*Shanghai Huanjing Gongbao*), www.envir.gov.cn

27 For instance, Shanghai is an experimental location for constructing an 'environmentally friendly city', supported by UNDP. Another project is dedicated to improving the quality of drinking water in Shanghai, supported by the World Bank. www.sepb.gov.cn

28 *Ibid.*

29 Available from *Shanghai Environmental Bulletin* (*Shanghai Huanjing Gongbao*) on Shanghai Environmental Protection Bureau's website, www.sepb.gov.cn

30 *Ibid.*

31 Interview with Ms Shao, program officer of GC, 11 January 2005. Interview with Mr Qian, previous participant in SGSF, 13 January 2005.

32 It is very difficult for grassroots NGOs to find a supervising institution that is required by registration.

33 Interview with Ms Fang, core member of Grassroots Community, 25 January 2005.

34 This is a government-organized, non-profit organization that politically belongs to the CCP. The league is organized according to the Party pattern. The Communist Youth League was also responsible for guiding the activities of the Young Pioneers (for children below the age of 14).

35 The WCS established an office in Shanghai in the late 1980s. It conducts education and research in protecting animal species.

36 For instance, Shoots and Roots, one international NGO committed to environmental education, cooperated with the local enterprises HP Inc. and Alcoa Inc. in its program in 2005. In the same year, the WCS collaborated with l'Oreal (Shanghai) to initiate an educational program for 'Concern on Yangtze, Protect Chinese River Dolphin'.

37 Currently the numbers of these two types of NGO are not available.

38 Statistics provided by the Shanghai Nomad Club—a web-based traveling group, which claims to have 15,200 registered members.

39 The GC was established in 2001; WWF Shanghai was formed around 2000; and the SGSF was established around 1998.

40 Available from website of the Grassroots Community. www.community.org.cn/en

41 Interview with Mr Liu, one of the main leaders of GC, 1 January 2005.

42 Interview with Mr Zhu, one of the main leaders of GC, 1 January 2005.

43 Green Power is an ENGO that commits to providing training to small grassroots NGOs. Interview with Mr Qu, leader of Green Power, 2 January 2005. Interview with Mr Yao, leader of Shanghai Wild Birds' Association, 20 May 2006.

44 *Ibid.*

45 Available from website of Grassroots Community. www.community.org.cn/en

46 Interview with Ms Ji, previous project officer of the 'environmental community' project in GC, 12 March 2005.

47 A propagating activity 'not to eat wild animals' was conducted in Lingguang Community, 15 November 2003, with around 500 inhabitants being educated.

48 Another propagating activity of 'genetically modified food' was conducted in Lingguang community, 15 February 2003, educating inhabitants as customers to avoid consuming genetically modified food.

49 This organization was also supported by the Canadian Center for Civil Society Development, available from Grassroots Community website. www.community.org.cn/en

50 No. 710 RC is one of the subordinate RCs of Zhijiang Xilu. It governs four inhabitants' communities: 591 Community, 710 Community, Liangguang Garden Community, and Guanglong Guangxing Community. Three classes of people live in this sub-district: poorer population in 591 Community; medium-income population in 710 Community; and higher-income population living in Lingguang Garden Community. Inhabitants who live in Zhijiang Xilu Community are mainly retired workers, who live a relatively poor life. On the whole, their concern about their environment is rather low.

51 Interview with Ms Ji, previous project officer of the 'environmental community' project in GC, 12 March 2005. Interview with Mr Mao, previous program officer of GC, 7 January 2005.

52 Interview with Mr Liang, inhabitant of No. 710 Community, 19 January 2005.

53 Interview with Mr Li, member of WWF Shanghai, 29 January 2005.

54 Interview with Mr Qu, president of Green Power Shanghai, 8 February 2005.

55 For instance, Urban Group of Shanghai Normal University won the National prize—the Earth Prize—for its research in 2002.

56 Interview with Professor Tao, supervising teacher of the Urban Group in Shanghai Normal University, 9 January 2005.

57 *Ibid.*

58 *Ibid.*

59 *Oriental Morning Press*, *Youth Daily* and *Wenhui News* provide special reports on this event. Available from internal introductory materials of SGSF.

60 The Municipal Youth League is the mother-in-law institution of SGSF.

61 Interview with Ms Yan, former member of SGSF from Shanghai Maritime University, 20 January 2005.

62 Interview with Mr Liu, former member of SGSF, 8 January 2005. Interview with Professor Tao, supervising teacher of the Urban Group in Shanghai Normal University, 9 January 2005.

63 'Problems that may affect young people and management work after China's participation of WTO', No. 36, available from website of Shanghai Youth League. www.why.com.cn/zuzhizhuanqu/yanjiu/canyue/cy2001.htm

64 On 17 April 2002, leaders from SGSF were invited to the meeting of preparation for the World Environment Day that was organized by Shanghai Environmental Education Center.

65 A letter was sent to the Mayor in November 2003.Information obtained from Professor Tao, supervising teacher of Urban Group of Shanghai Normal University, 9 January 2005.

66 One typical campaign was called 'Protect Mother River for Development of Eco-World Exposition in an Ecological City'. In this 2005 campaign, SGSF organized 13 university student groups to participate. Available from website of student environmental activists. www.gsean.org

67 WWF China is a branch of WWF International. It entered China in the 1980s, and was the first international ENGO to work in China. WWF China works closely with the Chinese government. WWF China provides virtual spaces for environmental activists in six cities, by cooperating with the website charged by *Renmin Daily*.

68 Registered individuals are regarded as members. Those who visit the website as regularly as twice or three times per week are defined as core members. Only the core members represent the voluntary group well.

69 Shi Zaishui, 'Shanghai wild birds protection association was established', *China Bird Watch*, 27 September 2004.

70 Interview with Mr Zhao, core member of SWBPA, 14 April 2005.
71 Interview with Mr Fang, core member of WWF Shanghai, 12 January 2005.
72 Interview with Mr Yang, core member of WWF Shanghai, 21 January 2005.
73 Because the internet is controlled by the government, radical ideas are still regulated by the political authorities concerned. But free speech can be made in anonymity.
74 A great number of individuals work in companies, where they are provided with a computer, and thus can access the internet when they are at work. University students can access the internet easily on campus, as most of them live on campus.
75 Professors of the Biology Department in East China Normal University and Shanghai Normal University were often contacted. Interview with Mr Zhao, core member of WWF Shanghai, 14 April 2005.
76 Interview with Mr Fang, core member of WWF Shanghai, 20 January 2005.
77 Interview with Mr Yao, core member of WWF Shanghai, 22 January 2005.
78 A second-grade social organization means that SWBPA will not be able to possess an independent bank account without the permission of its supervising institution.
79 Interview with Mr Yao, core member of WWF Shanghai, 20 May 2006.
80 *Ibid.*; interview with Mr Zhao, core member of SWBPA, 14 April 2005.
81 To be introduced in the final section of this chapter.
82 Interview with Ms Zhang, core member of WWF Shanghai, 20 April 2005.
83 Available from Yangpu district government official website. www.shyp.gov.cn/html/website/website/list/index.html
84 Interview with Mr Li, activist of the campaign, 8 January 2005.
85 Interview with Mr Yao, core member of WWF Shanghai, 4 April 2005.
86 For example, the teacher/advisor of Shanghai Normal University's environmental association did not allow students to get involved in this incident.
87 Available from the website of Shanhgai Nomad Club, www.hinomad.com; the website of Common Wealth Management Project, www.almsdeed.com.
88 Interview with Mr Zhao, core member of WWF Shanghai, 11 May 2005.
89 Shen, Ying, 'War on protecting wetland in cosmopolitan'; 'Exploring the wetland of the previous Jiangwan airport'; 'Visiting Jiangwan', *Southern Weekly*, 21 March 2003; Bo, Xiaobo, 'Exploring Jiangwan airport', *Wenhui Newspaper*, 21 May 2005.
90 Experts from the Shanghai Science Museum, the Youth and Education Center of Science and Technology, and scholars from Huadong Normal University Institute of Estuarine and Coastal Research participated in this meeting.
91 Available from the website of Shanghai Municipal People's Congress. www.shyp.gov.cn/html/yangpuPolity/yangpuPolity/list/index.html
92 Interview with Mr Fang, Mr Yao, and Mr Zhao, activists from WWF Shanghai, 12 January, 13 January and 14 April 20005, respectively. It is estimated that the security department of the municipal government was responsible for this.
93 Interview with Mr Fang, core member of WWF Shanghai, 16 May 2005.
94 Two institutes were mainly responsible for the EIA: Shanghai Environmental Impact Assessment Center, and Shanghai Urban Planning and Design Research Institute.
95 What is more, the result was doubtful to the environmentalists. According to them, the supportive scholars used to ally with environmentalists to oppose the construction. They were affiliated to government and the SNJI. Interview with Mr Li, main activist of WWF Shanghai, 10 January 2005. Interview with Ms Shen, journalist from *Nanfang Weekly*, 20 January 2005.
96 Tao Jian, 'What about ecology in Shanghai City's development?' *Liberation Daily*, 8 April 2003. Zhang Yi and Zheng Hon, 'Debate over Shanghai's last natural ecological region should be stopped', *International Finance Newspaper*, 28 March 2004.
97 Interview with Mr Li, core environmental activist in this campaign, 8 January 2005.
98 Interview with Qingfeng, core environmental activist, 23 January 2005.

99 Interview with Mr Fang and Mr Li, both core environmental activists in this campaign, 8 and 10 January 2005, respectively.

100 Estimated by members of GC. It is the number counted by average age of volunteers and committees leaders. The average age of volunteers is 28, and committee leaders are aged 30 on average. In reality, if all the members are counted, the real age on average should be older than 29, because a small portion of volunteers are retired people over 60.

101 Interview with Mr Guo, core member of GC, 20 May 2005. Interview with Mr Yao, core member of WWF Shanghai and SWBPA, 22 May 2005.

7 Comparison and conclusions

1 This has not been the case in, for instance, the Falun Gong, Taiwan, the Party, military developments, Tibet, and foreign affairs. The environment is, however, clearly indicated as a less sensitive issue, although at times reporting is also clearly regulated and restricted, such as on the Three Gorges Dam, the Harbin disaster, and spatial planning and land property rights.

2 But also the local culture of *guanxi* differs between the two cities, offering an alternative explanation of the differences in reliance on personal networks, as indicated below.

References

ACEF (2006) Report on environmental NGOs, internal report, All-China Environment Federation.

Alber, J. (1985) Modernisierung, neue spannungslinien und die politischen chancen der grünen, *Politische Vierteljahresschrift*, 267, 211–226.

Alexander, V. D. (1998) 'Environmental constraints and organizational strategies: complexity, conflict, and coping in the nonprofit sector', in W. W. Powell and E. S. Clemens (eds) *Private Action and the Public Good*, New Haven, CT: Yale University Press.

Alston, J. P. (1989) Wa guanxi, and inhwa: managerial principles in Japan, China, and Korea, *Business Horizons*, 3–4, 26–31.

Anheier, H. K. and Katz, H. (2005) Network approaches to global civil society, in H. K. Anheier, M. Glasius and M. Kaldor (eds) *Global Civil Society 2004/5*, London: Sage.

Bates, G. (2006) China's environmental challenge, testimony before the U.S.–China Economic and Security Review Commission hearing on major challenges facing the Chinese leadership. www.uscc.gov/hearings/2006hearings/hr06_02_02_03.php

Beach, M. (2002) Local Environment Management in China, *China Environment Series* 6, 21–31, Washington, DC: Woodrow Wilson Center.

Beach, M., Bleish, B. and Yang, S. (2006) The Role of Public Participation in Ecological Impact Assessment (EcIA) and Environmental Impact Assessment (EIA) in China, *China Environment Series* 8, 117–121, Washington, DC: Woodrow Wilson Center.

Bearman, P. (1993) *Relations into Rhetorics: Local Elite Social Structure in Norfolk, England, 1540–1640*, New Brunswick, NJ: Rutgers University Press.

Benford, R. D. (1994) Review of talking politics, by William A. Gamson, *American Journal of Sociology*, 99(4), 1103–1104.

Benford, R. D. and Snow, D. A. (2000) Framing processes and social movements: an overview and assessment, *Annual Review of Sociology*, 26, 611–639.

Bernstein, T. P. and Lü, X. (2003) *Taxation without Representation in Contemporary Rural China*, Port Chester, NY: Cambridge University Press.

Berscheid, E. (1985) Interpersonal attraction, in G. Lindzey and E. Aronson (eds) *Handbook of Social Psychology*, 3rd edn, New York: Random House.

Berscheid, E., Snyder, M. and Omoto, A. M. (1989) The relationship closeness inventory: assessing the closeness of interpersonal relationships, *Journal of Personality and Social Psychology*, 57, 792–807.

Bian, Y. (1997) Bringing strong ties back in: indirect ties, network bridges, and job searches in China, *American Sociological Review*, 62, 366–385.

—— (2002) Institutional holes and job mobility rrocesses: *guanxi* mechanism in China's emergent labor markets, in T. Gold, D. Guthrie and D. Wank (eds) *Social Connections in China: Institutions, Culture, and the Changing Nature of Guanxi*, Cambridge: Cambridge University Press.

Bianco, L. (2001) *Peasants Without the Party: Grass Roots Movements in Twentieth Century China*, London: M.E. Sharpe.

Boudreau, V. (1996) 'Northern theory, southern protest: opportunity structure analysis in cross–national perspective', *Mobilization*, 2, 175–189.

Brass, D. J. (1995) A social network perspective on human resources management, *Research in Personnel and Human Resources Management*, 13, 39–79.

Brass, D. J. and Burkhardt, M. E. (1993) Potential power and power use: an investigation of structure and behavior, *Academy of Management Journal*, 36, 441–470.

Brechin, S. R. and Kempton, W. (1994) Global environmentalism: a challenge to the postmateralism thesis?, *Social Science Quarterly*, 75(2), 245–269.

Brettell, A. (2000) Environmental non-governmental organizations in the people's republic of China: innocents in a co–opted environmental Movement, *Journal of Pacific Asia*, 6, 27–56.

—— (2008) 'Channeling dissent: the institutionalization of environmental complaint resolution', in P. Ho and R. Edmonds (eds) *China's Embedded Activism Opportunities and Constraints of a Social Movement*, London/New York: Routledge.

Broadbent, J. (2003) Movement in context: thick networks and Japanese environmental protest, in M. Diani and D. McAdam (eds) *Social Movement and Networks: Relational Approaches to Collective Action*, Oxford: Oxford University Press.

Buckley, L. (2006) Notes on Public Participation in Environmental Impact Assessments in China, *China Environment Series* 8, 112–116, Washington, DC: Woodrow Wilson Center.

Buechler, S. M. (1995) New social movement theories, *Sociological Quarterly*, 36, 441–464.

Burt, R. (1992) *Structural Holes*, Cambridge/London: Harvard University Press.

Cable, S. (1984) Professionalization in social movement organizations: a case study of Pennsylvanians for biblical morality, *Sociological Focus*, 17, 287–304.

Cai, Y. (2006) *State and Laid–off Workers in Reform China: The Silence and Collective Action of the Retrenched*, New York/Abingdon, Oxon: Routledge.

—— (2007) 'Civil resistance and rural of law in China: the case of defending home owners' rights', in Perry E. and Goldman, M. (eds) *Grassroots Politics in China*, Cambridge: Harvard University Press.

Cao, C. and Suttmeier, R. P. (2001) China's new scientific elite: distinguished young scientists, the research environment and hopes for Chinese science, *The China Quarterly*, 168, 960–984.

Carley, K. (1991) A theory of group stability, *American Sociology Review*, 56, 331–354.

Castells, M. (1996) *The Rise of the Network Society*, Oxford/New York: Blackwell.

—— (1997) *The Power of Identity*, Oxford/New York: Blackwell.

Chen, B. H. (2001) History of Shanghai Culture (*shanghai wenhua tongshi)*, Shanghai: Shanghai Artistic Press.

Chen, X. P. and Chen, C. C. (2004) On the intricacies of the Chinese guanxi: a process model of guanxi development, *Asia Pacific Journal of Management*, 21, 315–324.

Chung, C. (2006) Prospects for Developing Meaningful Public Participation in Environmental Decision-Making in China, *China Environment Series* 8, 138–141, Washington, DC: Woodrow Wilson Center.

Cohen, J. (1985) Strategy or identity: new theoretical paradigms and contemporary social movements, *Social Research*, 52(4), 664–716.

Coleman, J. S. (1973) *The Mathematics of Collective Action*, Chicago: Aldine.

—— (1977) 'Social action systems', in K. Szaniawski (ed.) *Problems of Formalization in the Social Sciences*, Wraclaw: Polskije Akademij Nauk.

Cotgrove, S. and Duff, A. (1980) Environmentalism, middle class radicalism and politics, *Sociological Review*, 28, 333–351.

Creswell, J. W. (1994) *Research Design: Qualitative and Quantitative Approaches*, Thousand Oaks, CA: Sage.

Cress, D. M. and Snow, D. A. (1996) Mobilization at the margins: resources, benefactors, and the viability of homeless social movement organizations, *American Sociological Review*, 61(6), 1089-1109.

Curtis, R. L. and Zurcher, L. A. (1973) Stable resources of protest movements: the multi-organizational field, *Social Forces*, 52, 53–61.

Dahl, R. A. (1958) 'A critique of the ruling elite model', *American Political Science Review*, 52, 463–469.

Dalton, R. J. and Kuechler, M. (eds) (1990) *Challenging the Political Order: New Social and Political Movements in Western Democracies*, New York: Oxford University Press.

Della Porta, D. (2005) 'Making the polis: social forums and democracy in the global justice movement', *Mobilization*, 10(1), 73–94.

Della Porta, D. and Diani, M. (2006) *Social Movements: An Introduction*, Oxford: Blackwell.

Deng, G. (2001) A preliminary analysis of the NGO survey, in M. Wang (ed.) *Research on NGOs in China* (*zhongguo NGO Yanjiu)*, Vol. 43, Beijing: UNCRD Research Report Series.

—— (2006) Two development paths of environmental NGOs. www.ngo.org.cn

Diani, M. (1990) The Italian ecology movement: from radicalism to moderation, in W. Rüdig (ed.) *Green Politics*, Edinburgh: Edinburgh University Press.

—— (1992) The concept of social movement, *Sociological Review*, 40, 1–25.

—— (1995) *Green Networks: A Structural Analysis of the Italian Environmental Movement*, Edinburgh: Edinburgh University Press.

—— (2000) Simmel to Rokkan and beyond: towards a network theory of (new) social movements , *European Journal of Social Theory*, 3(4), 387–406.

—— (2002) Network Analysis, in B. Klandermans and S. Staggenborg (eds) *Methods in Social Movement Research*, Minneapolis: University of Minnesota Press.

—— (2003) Leaders or brokers? positions and influence in social movement networks, in M. Diani and D. McAdam (eds) *Social Movement and Networks: Relational Approaches to Collective Action*, Oxford: Oxford University Press.

Diani, M. and Donati, P. R. (1999) Organizational change in western European environmental groups: a framework for analysis, *Environmental Politics*, 8, 13–34.

Duara, P. (1988) *Culture, Power and the State*, Stanford: Stanford University Press.

Duit, A. (2005) Understanding environmental performance of states: an institution-centered approach and some difficulties, from Göteborg University, QOG Institute. www.qog.pol.gu.se/working_papers/2005_7_Duit.pdf

Economy, E. C. (2003) China's Environmental Challenge, testimony before the Congressional Executive Commission on China Roundtable on the Environment. US Government Council on Foreign Relations. www.cfr.org/publication/5573/chinas_environmental_challenge.html

—— (2004) *The River Runs Black: The Environmental Challenge to China's Future*, Ithaca, NY: Cornell University Press.

—— (2005) China's environmental movement, testimony before the Congressional Executive Commission on China Roundtable on Environmental NGOs in China: Encouraging Action and Addressing Public Grievances, US Government Council on Foreign Relations. www.cfr.org/publication/7770

Eisinger, P. (1973) The conditions of protest behavior in American cities, *American Political Science Review*, 67, 11–28.

Emirbayer, M. and Goodwin, J. (1994) Network analysis, culture, and the problem of agency, *American Journal of Sociology*, 99(6), 1411–1454.

Fei, X. (1947) *From the Soil: The Foundations of Chinese Society*, 9th edn, Berkeley: University of California Press.

Fen, T. (2000) What is the Local Culture of Panjin? [*hewei panjin diyu wenhua*]. Panjin Achieve Library.

Friedman, D. and McAdam, D. (1992) Collective identity and activism: networks, choices, and the life of a social movement, in M. Aldon and C. M. Mueller (eds) *Frontiers in Social Movement Theory*, New Haven: Yale University Press.

Friends of Nature (2000) *1999 Survey on Environmental Reporting in Chinese Newspapers*, Beijing: Friends of Nature.

Fu, T. (2004) China's environmental NGOs—prosperity in participating environmental governance, in Environment and Development Research Center (eds), *China Environment Development Review*, Beijing: Social Sciences Documentation Publishing House.

Gale, R. P. (1986) 'Social movements and the state: the environmental movement, countermovement, and government agencies', *Sociological Perspectives*, 29, 202–240.

Gallagher, M. E. (2007) 'Hope for protection and hopeless choices': labor legal aid in the PRC, in E. J; Perry and M. Goldman (eds) *Grassroots Political Reform in Contemporary China*, Cambridge, MA: Harvard University Press

Gamson, W. A. (1975) *The Strategy of Social Protest*, Homewood, IL: Dorsey Press.

—— (1995) Constructing social protest, in H. Johnston and B. Klandermans (eds) *Social Movements and Culture*, London: UCL Press.

Gamson W. A. and Meyer, D. S. (1996) Framing political opportunity, in D. McAdam, J. D. McCarthy and M. N. Zald (eds) *Comparative Perspectives on Social Movements: Political Opportunities, Mobilizing Structure and Cultural Framing*, Cambridge, UK: Cambridge University Press.

Gamson, W. A. and Modigliani, A. (1989) Media discourse and public opinion on nuclear power: a constructionist approach, *American Journal of Sociology*, 95(1), 1–37.

Gamson, W. A. and Wolfsfeld, G. (1993) 'Movement and media as interacting systems', *Annals of the American Academy of Political and Social Science*, 528, 114–125.

Gamson, W. A., Fireman, B. and Rytina, S. (1982) *Encounters with Unjust Authorities*, Homewood, IL: Dorsey Press.

Ganz, M. (2000) 'Resources and resourcefulness: strategic capacity in the unionization of California agriculture, 1959–1966', *American Journal of Sociology*, 105, 1003–1062.

Gerhards, J. and Rucht, D. (1992) Mesomobilization: organizing and framing in two protest campaigns in west Germany, *American journal of sociology*, 98(3), 555-595.

Gold, T. (1985) After comradeship: personal relations in China since the cultural revolution, *The China Quarterly*, 104, 657–675.

Gold, T., Guthrie, D. and Wank, D. (eds) (2002) *Social Connections in China: Institutions, Culture, and the Changing Nature of Guanxi*, Cambridge, UK: Cambridge University Press.

Goode, W. J. (1959) The sociology of the family, in R. K. Merton, L. Broom and L. S. Cottrell Jr (eds) *Sociology Today*, New York: Basic Books.

Goodwin, J. and Jasper, J. M. (1999) Caught in a winding, snarling vine: the structural bias of political process theory, *Sociological Forum*, 14(1), 27–92.

Gottlieb, R. (1993) *The Transformation of the American Environmental Movement*, Covelo, CA: Island Press.

Gould, R. V. (1991) Multiple networks and mobilization in the Paris commune, 1871, *American Sociological Review*, 56, 716–729.

—— (2003) Why do networks matter? rationalist and structuralist interpretations, in M. Diani and D. McAdam (eds) *Social Movement and Networks: Relational Approaches to Collective Action*, Oxford: Oxford University Press.

Granovetter, M. S. (1973) The strength of weak ties, *American Journal of Sociology*, 6, 1360–1380.

Gudykunst, W. B. (1997) Cultural variability in communication: an introduction, *Communication Research*, 24(4), 327–348.

Gusfield, J. (1994) The reflexivity of social movements: collective behavior and mass society theory revisited, in E. Larana, H. Johnston and J. Gusfield (eds) *New Social Movements: From Ideology to Identity*, Philadelphia: Temple University Press.

Guthrie, D. (1998) Declining significance of guanxi in China's economic transition, *The China Quarterly*, 154, 254–282.

—— (1999) *Dragon in a Three–Piece Suit: The Emergence of Capitalism in China*, Princeton: Princeton University Press.

Haggelund, G. (2004) *Environment and Resettlement Politics in China: The Three Gorges Project*, Aldershot, UK: Ashgate.

Hampton, K. and Wellman, B. (2001) Long distance community in the network society, *American Behaviour Scientist*, 45, 477-496.

He, M. (2003) Political opportunity structure and study on social movement, Presentation on the Annual Conference of Taiwan Sociology Association, 29–30 November 2003. Taiwan National Sun Yat-Sen University Institutional Repository.

Ho, P. (2001) Greening without conflict? environmentalism, NGOs and civil society in China, *Development and Change*, 32, 893–921.

Ho, P. and Edmonds, R. (eds) (2008) *China's Embedded Activism Opportunities and Constraints of a Social Movement*, London/New York: Routledge.

Hofstede, G. (1983) Dimensions of national cultures in fifty countries and three regions, in J. Deregowsk (ed.) *Expectations in Cross-Cultural Psychology*, Lisse: Swets & Zeitlinger.

Hong, D. (2005) Environmental protection, public awareness and social coordination [*huanjing baohu, gongzhongyishi he shehuihexie*], in H. Zheng, L. Li and D. Hong (eds) *Report on China's Social Development* (zhongguo shehuifazhan yanjiu baogao 2005 [*zouxiang gengjia hexie de shehui*], Beijing: The People's University Press.

Howell, J. (2003) New directions in civil society: organizing around marginalized interests, in J. Howell (ed.) *Governance in China*, Lanham: Rowman & Littlefield.

Hsu, F. L. K. (1985a) *Culture and Self: Asian and Western Perspectives*, New York: Tavistock Publications.

—— (1985b) The self in cross-cultural perspective, in A. Marsella, G. DeVos and F. L. K. Hsu (eds) *Culture and Self*, New York: Tavistock Publications.

Hurlbert, J., Haines, V. A. and Beggs, J. J. (2000) Core networks and tie activation: what kind of routine networks allocate resources in nonroutine situations?, *American Sociological Review*, 65, 598–618.

Hwang, K. (1987) Face and favor: the Chinese power game, *American Journal of Sociology*, 92(4), 944–974.

—— (2000) Chinese relationalism: theoretical construction and methodological considerations, *Journal for Theory of Social Behavior*, 30(2), 155–178.

Inglehart R. (1971) The silent revolution in post-industrial societies, *American Political Science Review*, 65, 991–1017.

Inkeles, A. (1972) National character and modern political systems, in F. L. K. Hsu (ed.) *Psychological Anthropology*, Cambridge: Schenkman Books.

Institute of Environment and Development (2005) *Corporate Social Responsibility in China*, Beijing: Economic Science Press.

Jacobs, J. B. (1979) A preliminary model of particularistic tie in Chinese cultural setting, *The China Quarterly*, 78, 237–273.

Jahiel, A. R. (1997) The contradictory impact of reform on environmental protection, *The China Quarterly*, 149, 81–103.

—— (1998) The organization of environmental protection in China, *The China Quarterly*, 156, 757–787.

Jenkins, J. C. and Perrow, C. (1977) 'Insurgency of the powerless: farm worker movements (1946–1972)', *American Sociological Review*, 42, 249–267.

Ji, W. (2006) Rule of law and common trust—on reconstruction of order in the Chinese society, from a legal sociological perspective [*fazhi yu pubian xinren—guanyu zhongguo zhixuyuanli chonggou de fashehuixue shijiao*], *Journal of Law and Society*, 1, 166–191.

Jin, G. (1984) *Behind Historical Facts—Exploring Reasons for the Chinese Feudal Society's Strong Structure* [*zai lishi de biaoxiang beihou—dui zhongguo fengjian shehui chaowending jiegou de tansuo*], Taibei: Gufeng.

Katz, D. and Kahn, R. (1966) *The Social Psychology of Organizations*, New York: Wiley.

Kieschnick, J. (2003) *The Impact of Buddhism on Chinese Material Culture*, Princeton: Princeton University Press.

King, A. Y. (1991) Kuan-his and network building: a sociological interpretation, *Daedalus*, 120, 63–84.

Kitschelt, H. (1986) Political opportunity structures and political protest: anti-nuclear movements in four democracies, *British Journal of Political Science*, 16, 57–85.

Klandermans, B. (1984) Mobilization and participation: social–psychological expansions of resource mobilization theory, *American Sociological Review*, 49, 583–600.

—— (1989) Grievance interpretation and success expectations: the social construction of protest, *Social Behavior*, 4, 113–125.

—— (1991) New social movement and resource mobilization: the European and the American approach revisited, in D. Rucht (ed.) *Research on Social Movements: The State of the Art in Western Europe and the USA*, Boulder, CO: Westview Press.

Knoke, D. (1990) *Political Networks—The Structural Perspective*, New York: Cambridge University Press.

—— (1993) Networks of elite structure, *Sociological Methods and Research*, 22(1), 23–45.

Knup, E. (1997) Environmental NGOs in China: An Overview, *China Environment Series* 1, 9–15, Washington, DC: Woodrow Wilson Center.

Kriesi, H. (1987) Local mobilization for the people's petition of the Dutch peace movement, in B. Klandermans, H. Kriesi and S. Tarrow (eds) *From Structure to Action*, Greenwich, CT.: JAI Press.

—— (1988) The interdependence of structure and action: some reflections on the state of the art, in B. Klandermans, H. Kriesi and S. Tarrow (eds) *From Structure to Action*, Greenwich, CT: JAI Press.

—— (1991) *The Political Opportunity Structure of New Social Movements: Its Impact on Their Mobilization*, Berlin: WZB.

—— (1992) New social movements and political opportunities in western Europe, *European Journal of Political Research*, 22(2), 219–244.

—— (1995) *Social Movements in Western Europe: A Comparative Analysis*, Minneapolis: University of Minnesota Press.

—— (1996) The organizational structure of new social movements in a political context, in D. McAdam, J. D. McCarthy and M. N. Zald (eds) *Comparative Perspectives on Social Movements: Political Opportunities, Mobilizing Structure and Cultural Framings*, Cambridge: Cambridge University Press.

—— (2004) Political context and opportunity, in D. A. Snow, S. A. Soule and H. Kriesi (eds) *The Blackwell Companion to Social Movements*, Oxford: Blackwell.

Koopmans, R. (1993) The dynamics of protest waves: West Germany, 1965 to 1989', *American Sociological Review*, 58, 637–658.

Kuhn, P. (1995) Ideas behind China's modern state, *Harvard Journal of Asiatic Studies*, 55(2), 295–337.

Landes, D. S. (1998) *The Wealth and Poverty of Nations*, New York: W.W. Norton & Co.

Laumann, E. O, Knoke, D. and Kim, Y. H. (1985) An organizational approach to state policy formation: a comparative study of energy and health domains, *American Sociological Review*, 50, 1–19.

Lee, C. (2007) *Against the Law: Labour Protests in China's Rustbelt and Sunbelt*, Berkeley: University of California Press.

Lee, S. (2003) More Players on the Stage: New Trends in Shanghai's Water Pollution Control Policies, *China Environment Series* 6, 110–115, Washington, DC: Woodrow Wilson Center.

—— (2006) *Water and Development in China: The Political Economy of Shanghai Water Policy*, Singapore/London: World Scientific.

Lee, Y. (2005) Public environmental consciousness in China: early empirical evidence, in D. A. Kristen (ed.) *China's Environment and the Challenge of Sustainable Development*, Armonk, NY: M.E. Sharpe.

Lee, Y. and So, A. Y. (eds) (1999) *Asia's Environmental Movements: Comparative Perspectives*, Armonk, NY: M.E. Sharpe.

Lester, R. (1988) *Environmental Policy in China*, Bloomington: Indiana University Press.

Levenson, J. (1964) *Confucian China and its Modern Fate*, Berkeley: University of California Press.

Li, C. (1996) Rediscovering urban subcultures: the contrast between Shanghai and Beijing, *The China Journal*, 36, 139–153.

Li, L. and O'Brien, K. (2008) Protest leadership in rural China, *The China Quarterly*, 193, 1–23.

Li, M. (1990) Moral education in the people's republic of China, *Journal of Moral Education*, 19, 159–171.

Lieberthal, K. (1984) Domestic politics and foreign policy, in H. Harding (ed.) *China's Foreign Policy in the 1980s*, New Haven, CT: Yale University Press.

—— (1997) China's Governing System and its Impact on Environmental Policy Implementation, *China Environmental Series* 1, 3–8, Washington, DC: Woodrow Wilson Center.

Lieberthal, K. and Lampton, D. M. (eds) (1992) *Bureaucracy, Politics, and Decision Making in Post–Mao China*, Berkeley: University of California Press.

Lin, N. (2001) *Social Capital: A Theory of Social Structure and Action*, New York: Cambridge University Press.

Lin, N. and Bian, Y. (1991) Getting ahead in urban China, *American Journal of Sociology*, 97, 657–688.

Lincoln, J. R. (1978) Community structure and industrial conflict, *American Sociological Review*, 43, 199–220.

Lindley, C. (1975) Changing policy management responsibilities of local legislative bodies, *Public Administration Review*, 35, 794–797.

Liu, X. (1992) *This Generation's Fear and Love*, Beijing: Sanlian Publishing House.

Lo, C., Yip, W. P. W. and Cheung, K. C. (2000) The regulatory style of environmental governance in China: the case of EIA regulation in Shanghai, *Public Administration and Development*, 20, 305–318.

Lu, H. (2001) *A Survey of the Current Conditions of University Environmental Associations in China* [*zhongguo gaoxiao huanbao shetuan xianzhuang diaocha*], Chengdu: Sichuan University.

Ma, A. (2006) *Non-governmental Organizations in Contemporary China: Paving the Way to Civil Society?* London: Routledge.

Ma, X. and Ortolano, L. (2000) *Environmental Regulation in China: Institutions, Enforcement and Compliance*, Lanham: Rowman & Littlefield.

Martens, S. (2006) Public participation with Chinese characteristics: citizen consumers in China's environmental management, *Environmental Politics*, 15(2), 211–230.

Marwell, G., Oliver, P. and Prahl, R. (1988) Social networks and collective action: a theory of the critical mass, *American Journal of Sociology*, 94(3): 502–534.

McAdam, D. (1986) Recruitment to high risk activism: the case of freedom summer, *American Journal of Sociology*, 92, 64–90.

—— (1996) The framing function of movement tactics: strategic dramaturgy in the American civil rights movement, in D. McAdam, J. D. McCarthy and M. N Zald (eds) *Comparative Perspectives on Social Movements: Political Opportunities, Mobilizing Structure and Cultural Framing*, Cambridge, UK: Cambridge University Press.

—— (2003) Beyond structural analysis: toward a more dynamic understanding of social movements, in M. Diani and D. McAdam (eds) *Social Movement and Networks: Relational Approaches to Collective Action*, Oxford: Oxford University Press.

McAdam, D. and Fernandez, R. (1990) Microstructural bases of recruitment to social movements, in. L. Kriesberg (ed.) *Research in Social Movements, Conflict and Change* 12, Greenwich, CT: JAI Press.

McAdam, D. and Paulsen, R. (1993) Specifying the relationship between social ties and activism, *American Journal of Sociology*, 99(3), 640–667.

McAdam, D., Sampson, R. J., Weffer, S. and MacIndoe, H. (2005) There will be fighting in the streets: the distorting lens of social movement theory, *Mobilization*, 10(1), 1–18.

McCaffrey, D. and Keys, J. (2000) Competitive framing processes in the abortion debate: polarization–vilification, frame saving, and frame debunking, *Sociological Quarterly*, 41, 41–61.

McCarthy, J. D. and Wolfson, M. (1992) Consensus movements, conflict movements, and the cooptation of civic and state infrastructure, in M. Aldon and C. M. Mueller (eds) *Frontiers in Social Movement Theory*, New Haven: Yale University Press.

McCarthy, D. J. and Zald, M. N. (1973) The Trend of Social Movement in America: Professionalization and Resource Mobilization, Morristown, NJ: General Learning Press.

—— (1977) Resource mobilization and social movements: a partial theory, *American Journal of Sociology*, 82, 1212–1241.

—— (1988) Social movements, in N. J.Smelser (ed.) *Handbook of Sociology*, Newbury Park, CA: Sage.
McCarthy, J. D., Wolfson, M., Baker, D. P. and Mosakawski, E. (1973) *The Trend of Social Movement in America: Professionalization and Resource Mobilization*, Morristown, NJ: General Learning Press.
—— (1977) Resource mobilization and social movements: a partial theory, *American Journal of Sociology*, 82, 1212–1241.
Melucci, A. (1989) *Nomads of the Present: Social Movements and Individual Needs in Contemporary Society*, Philadelphia: Temple University Press.
—— (1996) *Challenging Codes: Collective Action in the Information Age*, Cambridge/New York: Cambridge University Press.
Metzger, T. (2001) The western concept of the civil society in the context of Chinese history, in S. Kaviraj and S. Khilnani (eds) *Civil Society: History and Possibilities*, Cambridge, UK: Cambridge University Press.
Meyer, D. S. and Minkoff, D. C. (2004) Conceptualizing political opportunity, *Social Forces*, 82(4), 1457–1492.
Minkoff, D. C. (1994) From service provision to institutional advocacy: the shifting legitimacy of organizational forms, *Social Forces*, 72(4), 943–969.
—— (1997) 'The sequencing of social movements', *American Sociological Review*, 62, 779–799.
—— (1999) Bending with the wind: organizational change in American women's and minority organizations, *American Journal of Sociology*, 104, 16661703.
—— (2001) Social movement politics and organizations, in J. R. Blau (ed.) *Blackwell Handbook of Sociology*, London: Blackwell.
Mueller, C. (1994) Conflict networks and the origins of women's liberation, in E. Larana, H. Johnston and J. Gusfield (eds) *New Social Movements: From Ideology to Identity*, Philadelphia: Temple University Press.
Minsheng Bank (2005) Survey on Chinese individuals' environmental awareness, www.most.gov.cn/gnwkjdt/200606/t20060605_33766.htm
Mol, A. P. J. (2006) Environment and modernity in transitional China: frontiers of ecological modernization, *Development and Change*, 37(1), 29–56.
Mol, A. P. J. and Carter, N. T. (2006) China's environmental governance in transition, *Environmental Politics*, 15(2), 149–170.
Moore, A. and Warren, A. (2006) Legal Advocacy in Environmental Public Participation in China: Raising the Stakes and Strengthening Stakeholders, *China Environment Series* 8, 3–26, Washington, DC: Woodrow Wilson Center.
Morton, K. (2008) 'Transnational advocacy at the grassroots in China: potential benefits and risks', in P. Ho and R. Edmonds (eds) *China's Embedded Activism Opportunities and Constraints of a Social Movement*, London/New York: Routledge.
Morris, A. D. and Staggenborg, S. (2004) Leadership in social movements, in D. A. Snow, S. A. Soule and H. Kriesi (eds) *The Blackwell Companion to Social Movements*, Oxford: Blackwell.
Neidhardt, F. and Rucht, D. (1991) The analysis of social movements: the state of the art and some perspectives for further research, in D. Rucht (ed.) *Research on Social Movements: The State of the Art in Western Europe and the USA*, Frankfurt/M and Boulder, CO: Campus and Westview Press.
Noakes, J. A. and Hank, J. (2005) Frames of protest: a road map to a perspective, in J. Hank and J. A. Noakes (eds) *Frames of Protest*, Lanham: Rowman & Littlefield.

North, D. (1990) *Institutions, Institutional Change, and Economic Performance*, Cambridge: Cambridge University Press.

Oberschall, A. (1973) *Social Conflict and Social Movements*, Englewood Cliffs, NJ: Prentice Hall.

—— (1980) Loosely structured collective conflict, in L. Kriesberg (eds) *Research in Social Movements, Conflict and Change*, Greenwich, CN: JAI Press.

O'Brien, K. (2002) Collective action in the Chinese countryside, *The China Journal*, 48, 139–54

O'Brien, K. and Li, L. (2004) Suing the local state: administrative litigation in rural China, *The China Journal*, 51, 76–96.

—— (2006) *Rightful Resistence in Rural China*, New York/ Cambridge: Cambridge University Press.

Office of Implementing Agenda 21 (1999) *China's Agenda 21—Actions in Shanghai*, unpublished, Office of Implementing Agenda 21.

Ohlemacher, T. (1996) Bridging people and protest: social relays of protest groups against low–flying military jets in West Germany, *Social Problems*, 43, 197–218.

Oliver, P. E., Cadena-Roa, J. and Strawn, K. D. (2003) Emerging trends in the study of protest and social movements, in B. A. Dobratz, T. Buzzell and L. K. Waldner (eds) *Research in Political Sociology*, Vol. 11, Stanford, CT: JAI Press.

Olzak, S. and Ryo, E. (2007) Organizational diversity, vitality, and outcomes in the African–American civil rights movement, *Social Forces*, 85, 1561–1592.

Pan, Y. (2006) Environmental protection and social equality. www.greengrants.org.cn

Passy, F. (2003) Social networks matter. But how?, in M. Diani and D. McAdam (eds) *Social Movement and Networks: Relational Approaches to Collective Action*, Oxford: Oxford University Press.

Perry, E. J. (2001) *Challenging the Mandate of Heaven: Social Protest and State Power in China*, Armonk, NY: M.E: Sharpe.

Pickerill, J. (2001) Weaving a green web: environmental protest and computer mediated communication in Britain, in F. Webster (eds) *Culture and Politics in the Information Age*, London: Routledge.

Putnam, R. D. (1993) *Making Democracy Work: Civic Traditions in Modern Italy*, Princeton, NJ: Princeton University Press.

Pye, L. W. (1981) *The Mandarin and the Cadre—China's Political Cultures*, Ann Arbor: University of Michigan Press.

—— (1986) On Chinese pragmatism in the 1980s, *The China Quarterly*, 106, 207–223.

Pye, L. W. and Pye, M. W. (1985) *Asian Power and Politics: The Cultural Dimensions of Authority*, Cambridge, MA: Harvard University Press.

Qian, T. (2006) NGO merger counters trend towards fragmentation, *China Development Brief*, 5. www.chinadevelopmentbrief.com/node/614

Research Team on Sustainable Development (2005) *Report on China's Sustainable Development* [*zhongguo kechixu fazhan zhanlue baogao*], Beijing: Science Press.

Rosenstone, S. and Hansen, J. M. (1993) *Mobilization, Participation, and Democracy in America*, New York: Macmillan.

Rosenthal, N., Fingrutd, M., Ethier, M., Karant, R. and McDonald, D. (1985) Social movements and network analysis: a case study of nineteenth-century women's reform in New York state, *American Journal of Sociology*, 90(5), 1022–1054.

Royo, A. G. (2000) The power of networking: building force to navigate cross-scale turbulence where solo efforts fail, in J. B. Alcorn and A. G. Royo (eds) *Indigenous*

Social Movements and Ecological Resilience: Lessons From Dayak of Indonesia, Washington, DC: World Wildlife Fund.
Rubin, Z. (1973) *Liking and Loving: An Invitation to Social Psychology*, New York: Holt, Rinehart & Winson.
Rucht, D. (1990) The strategies and action repertoire of new movements, in R. J. Dalton and M. Kuechler (eds) *Challenging the Political Order: New Social Movements in Western Democracies*, Cambridge: Polity Press.
—— (1995) The impact of anti-nuclear power movements in international comparison, in M. Bauer (ed.) *Resistance to New Technology*, Cambridge: Cambridge University Press.
—— (1996) The impact of national contexts on social movement structures: a cross-movement and cross-national comparison, in D. McAdam, J. McCarthy and M. Zald (eds) *Comparative Perspectives on Social Movements*, Cambridge: Cambridge University Press.
Ruzza, C. E. and Schmidtke, O. (1993) Roots of success of the lega lombarda, *West European Politics*, 16(2), 1–23.
Saich, T. (2000) Negotiating the state: the development of social organizations in China, *The China Quarterly*, 3, 125–141.
Salamon, L. M. and Anheier, H. K. (1997) Toward a common classfication, in L. M. Salamon and H. K. Anheier (eds) *Defining the Nonprofit Sector: A Cross-National Analysis*, Manchester: Manchester University Press.
Schneider, C. (2005) Political opportunities and framing Puerto Rican identity in New York City, in J. Hank and J. A. Noakes (eds) *Frames of Protest*, Lanham: Rowman & Littlefield.
Schlosberg, D. (1999) Networks and mobile arrangements: political innovation in the U.S. environmental justice movement, *Environmental Politics*, 6(1), 122–148.
Schwartz, J. (2003) The impact of state capacity on enforcement of environmental policies: the case of China, *Journal of Environment & Development*, 12(1), 50–81.
—— (2004) Environmental NGOs in China: roles and limits, *Pacific Affairs*, 77(1), 28–49.
Shambaugh, D. (ed.) (2000) *The Modern Chinese State*, Cambridge, UK: Cambridge University Press.
Shanghai Chorography Editing Office (1997) *Shanghai Chorography of Tax and Finance*, Shanghai: Social Sciences Press.
Shanghai Municipal Statistics Bureau (2006) *Shanghai Statistical Yearbook*, Beijing: China Statistics Press.
Shen, H. (2002) Directing in order—Shanghai's NGO administration, unpublished, Beijing: Tsinghua University, NGO Research Center.
Shi, T. (1997) *Political Participation in Beijing*, Cambridge, MA: Harvard University Press.
—— (2000) Cultural values and democracy in the people's republic of China, *The China Quarterly*, 162, 540–569.
Shue, V. (1988) *The Reach of the State: Sketches of the Chinese Body Politic*, Stanford: Stanford University Press.
Simmel, G. (1955) *Conflict and the Web of Group Affiliations* [1908], trans. R. Bendix, New York: Free Press.
Smith, J. and Wiest, D. (2003) The uneven geography of global civil society: national and global influences on transnational association, unpublished, State University of New York at Stony Brook.
Snow, D. A. (2000) Framing processes and social movements: an overview and assessment, *Annual Review of Sociology*, 26, 611–629.

—— (2004) Framing processes, ideology and discursive fields, in D. A. Snow, S. A. Soule and H. Kriesi (eds) *The Blackwell Companion to Social Movements*, Oxford: Blackwell.

Snow, D.A. and Benford R. D. (1988) 'Ideology, frame resonance and participant mobilization, *International Social Movement Research*, 1, 197-219.

Snow, D. A. and Trom, D. (2002) The case study and the study of social movements, in B. Klandermans and S. Staggenborg (eds) *Methods of Social Movement Research*, Minneapolis: University of Minnesota Press.

Snow, D. A., Zurcher L. A. and Ekland-Olson, S. (1980) Social networks and social movements: a microstructural approach to differential recruitment, *America Sociological Review*, 45, 787–801.

Snow, D. A., Rochford, B. E. Jr, Worden, S. K. and Benford, R. D. (1986) Frame alignment processes, micromobilization, and movement participation, *American Sociological Review*, 51(4), 464–481.

So, A. Y. and Lee, Y. (1999) *Asia's Environmental Movements: Comparative Perspectives.* Armonk, NY: M.E. Sharpe.

Soule, S. A., McAdam, D., McCarthy, J. and Su, Y. (1999) Protest events: cause or consequence of state action? The U.S. women's movement and federal congressional activities, 1956–1979, *Mobilization*, 4(2), 239–255.

Staggenborg, S. (1986) Coalition work in the pro-choice movement: organizational and environmental opportunities and obstacles, *Social Problems*, 33(5), 374–390.

—— (1988) The consequences of professionalization and formalization in the pro-choice movement, *American Sociological Review*, 53(8), 585–606.

Stalley, P. and Yang, D. (2006) An emerging environmental movement in China?, *China Quarterly*, 333–356.

Sullivan, R. L. (1994) The three gorges dam and the Chinese polity, in Q. Dai, *Yangtze! Yangtze.* www.threegorgesprobe.org/tgp/books.html

Sun, L. (1996) Personal relationships, social connections and social structure [*guanxi, shehui guanxi he shehui jiegou*], *Sociological Studies*, 5, 27–54.

Sun, Y. and Zhao, D. (2008) State–society relations and environmental campaigns, in K. O Brien (ed.) *Popular Protest in China*, Cambridge: Harvard University Press.

Tang, S., Lo, C. W., Cheung, K. and Lo, J. M. (1997) Institutional constraints on environmental management in urban China: environmental impact assessment in Guangzhou and Shanghai, *The China Quarterly*, 152(12), 863–874.

Tang, X. (2004) *Wrong, Wrong, Wrong—Tang Xiyang's Green Thoughts and Hundred's Comments* [*cuo, cuo cuo—tang xiyang lvse hensi yu baijia pingdian*], Shenyang: Shenyang Press.

Tarrow, S. (1983) *Struggling to Reform: Social Movements and Policy Change During Cycles of Protest*, Western Societies Program, Occasional Paper 15, Ithaca, NY: Center for International Studies, Cornell University.

—— (1988) National politics and collective action: recent theory and research in western Europe and the United States, *Annual Review of Sociology*, 14, 421–440.

—— (1992) Mentalities, political cultures, and collective action frames, in A. D. Morris and C. M. Mueller (eds) *Frontiers in Social Movement Theory*, New Haven: Yale University Press.

—— (1994) *Power in Movement: Social Movements, Collective Action and Politics*, Cambridge: Cambridge University Press.

—— (1996) Social movements in contentious politics: a review article, *American Political Science Review*, 90, 874–883.

—— (1998) *Power in Movement: Social Movements and Contentious Politics*, Cambridge, UK: Cambridge University Press.
Tilly, C. (1993) Contentious repertoires in Great Britain, 1758–1834, *Social Science History*, 17, 253–280.
—— (1995) *Popular Contention in Britain, 1758–1834*, Cambridge, MA: Harvard University Press.
—— (1998) Social movements and (all sorts of) other political interactions—local, national, and international—including identities, *Theory and Society*, 27, 453–480.
(2004) *Social Movements, 1768–2004*, Boulder, CO: Paradigm Press.
Tranvik, T. (2004) Surfing for online connectedness: is the internet helping to end civic engagement? in S. Prakash and P. Selle (eds) *Investigating social capital*, London: Sage.
Turner, J. (2003) Clearing the air: human rights and the legal dimensions of China's environmental dilemma, Congressional/Executive Commission on China Issues Roundtable. www.cecc.gov/pages/roundtables/index.php
—— (2004) Small government, big and green society, *Harvard Asia Quarterly*, 2. www.asiaquarterly.com/content/view/145/40/
Unger, J. (2002) *The Transformation of Rural China*, Armonk, NY: M. E. Sharpe.
Useem, M. (1979) The social organization of the American business elite and participation of corporate directors in the governance of American institutions, *American Sociological Review*, 44, 553–572.
Van der Heijden, H. A. (1997) Political opportunity structure and the institutionalisation of the environmental movement, *Environmental Politics*, 4(1), 25–50.
—— (2002) Dutch environmentalism in the 1990s, *Environmental Politics*, 11(4), 120–130.
—— (2006) Globalization, environmental movements and international political opportunity structures, *Organization and Environment*, 19(1), 28–45.
Verschuren, P. (2003) Case study as a research strategy: some ambiguities and opportunities, *International Journal of Social Research Methodology*, 6(2), 121–129.
Verschuren, P. and Doorewaard, H. (1999) *Designing a Research Project*, Utrecht: Lemma.
Walder, A. (1986) *Communist Neo-Traditionalism; Work and Authority in Chinese Industry*, Berkeley: University of California Press.
Walgrave, S. and Manssens, J. (2005) Mobilizing the white march: media frames as alternatives to movement organizations, in J. Hank and J. A. Noakes (eds) *Frames of Protest*, Lanham: Rowman & Littlefield.
Walgrave, S. and Rucht, D. (eds) (2007) *Protest Politics, Antiwar Mobilization in Advanced Industrial Democracies*, Minneapolis: University of Minnesota Press.
Walker, E. T. and McCarthy, J. D. (2007) The influence of organizational structure, membership composition and resources on the survival of poor people's social movement organizations, in M. Chen, R. Jhabvala, R. Kanbur and C. Richards (eds) *Membership Based Organizations of the Poor*, London/New York: Routledge.
Wang, A. (2005) The downside of growth: law, policy and China's environmental crises, *Perspective*, 2(2). www.oycf.org/oycfold/httpdocs/Perspectives2/8_103100/downside_of_growth.htm
Wang, J. and Tong, X. (2005) *Sustaining Urban Growth through Innovative Capacity: Beijing and Shanghai in Comparison*, World Bank Policy Research Working Paper. http://openlibrary.org/b/OL3477558M
Wang, L., Zhang, R. and Kong, F. (2004) Review of industrial environmental management in China. www.va-china.com/va/review%20IEM%20China%20(Eng).pdf
Wang, M. (ed.) (2000) *Case Study on China's NGOs*, Beijing: Tsinghua University Press.

Wang, S. (1997) *Bottom Line of Decentralization* [*fenquan de dixian*], Beijing: China Planning Publishing House.

Wang, S. and He, J. (2004) The associational revolution in China—a comprehensive review on Chinese association behavior [*zhongguoren de shetuan geming*], *Zhejiang Xuekan*, 11, 71–77.

Wang, Y. (1987) Face and debt of emotions—the Chinese art of power [*renqing shengsuo, mianzi gongfu—zhongguoren de quanmou zhidao*], *Journal of Teacher Zhang*, 1, 11–15.

Wank, D. (1999) *Commodifying Communism: Business, Trust, and Politics in a Chinese City*, New York: Cambridge University Press.

—— (2003) Business–state clientelism in China, in T. Gold, D. Guthrie and D. Wank (eds) *Social Connections in China: Institutions, Culture, and the Changing Nature of Guanxi*, Cambridge, UK: Cambridge University Press.

Wasserman, S. and Faust, K. (1994) *Social Network Analysis: Methods and Applications*, Cambridge: Cambridge University Press.

Wellman, B. (1983) Network analysis: some basic principles, *Sociological Theory*, 1, 155–200.

Wellman, B. and Berkowitz, S. D. (eds) (1988) *Social Structures: A Network Approach*, Cambridge: Cambridge University Press.

Wellman, B., Chen, W. and Dong, W. (2002), Networking guanxi, in T. Gold, D. Guthrie and D. Wank (eds) *Social Connections in China: Institutions, Culture, and the Changing Nature of Guanxi*, Cambridge/New York: Cambridge University Press.

Wen, B. (2003) Weaving a Green Web: The Internet and Environmental Activism in China, *China Environment Series* 6, 89–93, Washington, DC: Woodrow Wilson Center.

Wen, X. (2006) *Review on HuXiang Culture* [*huxiangwenhua gujintan*], Changsha: Hunan People's Publishing House.

White, G., Howell, J. and Shang, X. (1996) *In Search of Civil Society: Market Reform and Social Change in Contemporary China*, Oxford: Clarendon Press.

White Paper (2006) Environmental protection in China (1996–2005). www.china.com.cn/english/MATERIAL/170257.htm

World Bank (1996) *The World Bank Participation Sourcebook*, Washington DC: World Bank.

Wu, F. (2002a) New Partners or Old Brothers? GONGOs in Transitional Environmental Advocacy in China, *China Environment Series* 5, 45–58, Washington, DC: Woodrow Wilson Center.

—— (2002b) Shanghai Greenies, *China Environment Series* 5, 99–101, Washington, DC: Woodrow Wilson Center.

Xi, X. and Xu, Q. (eds) (1999) *A Survey on China's Environmental Consciousness* [*zhongguo gongzhong huanjing yishi diaocha*], Beijing: China Environmental Science Press.

Xia, G. (2000) *Innovation of Environmental Policies* [*huanjing zhengce chuangxin*], Beijing: China Environmental Science Publishing.

Xia, M. (2000) Political contestation and the emergence of the provincial people's congresses as power players in Chinese politics: a network explanation, *Journal of Contemporary China*, 9(24), 185–214.

Xiangfan Publicity Dept (2003) *South–North Diversion—Rebuilding Ecology of Xiangfan* [*nanshuibeidiao—chongjian xiangfan shengtan*], unpublished, Xiangfan.

Xu, B. and Sun, S. (eds) (1996) *Northeastern Culture—Agricultural Culture among Mountains and Rivers* [*dongbei wenhua—baishan heishui zhong de nongye wenming*], Hong Kong: Commercial Press.

Yale Center and CIESIN (2006) Pilot 2006 environmental performance index, Yale Center for Environmental Law and Policy, and Center for International Earth Science Information Network. www.yale.edu/epi

Yan, C. (2005) The historical characteristic of Beijing culture [*Beijing wenhua de lishi tedian*], in Research Report Team (eds) *The Cultural Developmental Report of Beijing* [*Beijing wenhua fazhan baogao*], Beijing: Tongxin Press.

Yang, G. (2004) Global environmentalism hits China, *Yale Global*, 4. http://yaleglobal.yale.edu/display.article?id=3250

—— (2005a) Environmental NGOs and institutional dynamics in China, *The China Quarterly*, 181, 3:46–66.

—— (2005b) Information technology and grassroots democracy: a case study of environmental activism in China, in C. Nieman (ed.) *Democracy and Globalization*, Kent, OH: Kent State University Press.

—— (2007) How do Chinese civic associations respond to the internet: findings from a survey, *The China Quarterly*, 189, 122–143.

Yang, M. (1994) *Gifts, Favors, and Banquets: The Art of Social Relationships in China*, Ithaca, NY: Cornell University Press.

Yang, X. (2004) *Report on Social Development in Shanghai* [*Shanghai shehui fazhan lanpishu*], Shanghai: Social Science Press.

Yin, H. (2002) *Reappraisal of Cultural Change in Modern China* [*zhongguo wenhua de zhanwang*], Shanghai: Sanlian Shudian.

Yin, R. K. (1984) *Case Study Research: Design and Methods*, Newbury Park, CA: Sage.

Yuan, B. and Shen, J. (1998) Moral values held by early adolescents in Taiwan and Mainland China, *Journal of Moral Education*, 27(2), 191–208.

Yuan, F. (1999) A survey on Chinese residents environmental consciousness in Beijing and Shanghai [*zhongguo shimin de huanjing yishi diaocha: Beijing he Shanghai*], in X. Xi and Q. Xu (eds) *A Survey on China s Environmental Consciousness* [*zhongguo gongzhong huanjing yishi diaocha*], Beijing: China Environmental Science Press.

Yum, O. J. (1988) The impact of confucianism on interpersonal relationships and communication patterns in East Asia, *Communication Monographs*, 55, 374–388.

Young, N. (2001) Searching for civil society, in N. Young (ed.) *250 NGOs in China: Civil Society in the Making*, Beijing: China Development Brief:

Zald, M (1991) The continuing vitality of resource mobilization theory, in D. Rucht (ed.) *Research on Social Movements: The State of the Art in Western Europe and the USA*, Boulder, CO: Westview Press.

Zald, M. and McCarthy, J. D. (eds) (1987) *Social Movements in an Organizational Society*, New Brunswick, NJ: Transaction Books.

Zdravomyslova, E. (1996) Problems of becoming a housewife, in A. Rotkirch and E. Haavio-Mannila (eds) *Women's Voices in Russia Today*, Aldershot: Dartmouth.

Zhang, C. (2001) The financial resources of private enterprises—investigation on four provinces/municipals [*minying qiye rongzi qudao you duokuan—xishengshi minying qiye rongzi qingkuang diaoyan*], *New Economy Weekly*, 9, 12–17.

Zhang, J. (2000) *Problems of Rural Level Governance in China* [*jiceng zhengquan: xiangcun zhidu zhuwenti*], Hangzhou: Zhejiang People's Press.

Zhang, K. and Xia, G. (2000) New developmental pattern and the innovation of Chinese environmental policies [*xinfazhan fangshi yu zhongguo huanjing zhengce chuangxin*], *Dynamic of Environmental Science*, 1, 1–5.

Zhang, M. (2002) *Measuring Urban Sustainability in China*, Amsterdam: Thela Thesis.

Zhao, D. (2005) A critical analysis on the development of the western theories of social movements and revolutions: a Chinese synthesis, *Sociological Studies*, 115, 168–209.

Zhao, X. (2004) Strategies NGOs utilize against government: a preliminary investigation [*zhongguo NGO dui zhengfu de celue: yige chubu kaocha*], *Open Times*, 6, 5–23.

Zheng, Y. (2001) Introduction, in Environment and Development Research Institute *China Environment and Development Review*, Beijing: Social Sciences Documentation Publishing House.

Zhou, X. (2004) The rising environmental organizations in Shanghai [*zhengzai jueqi de shanghai huanbao zuzhi*]. www.greengrants.org.cn

Zhu, J. (2004) Grassroots NGOs and Chinese civil society's development, *Open Times*, 6, 36–48.

Zhu, J. and Ho, P. (2008) Not against the state, just protect the residents' interests: an urban movement in a Shanghai neighbourhood, in P. Ho and R. Edmonds (eds) *China's Embedded Activism Opportunities and Constraints of a Social Movement*, London/New York: Routledge.

Zusman, E. and Turner, J. (2005) Beyond the bureaucracy, changing China's policy-making environment, in D. A. Kristen (ed.) *China's Environment and the Challenge of Sustainable Development*, Armonk, NY: M. E. Sharpe.

Index

eBooks

eBooks – at www.eBookstore.tandf.co.uk

A library at your fingertips!

eBooks are electronic versions of printed books. You can store them on your PC/laptop or browse them online.

They have advantages for anyone needing rapid access to a wide variety of published, copyright information.

eBooks can help your research by enabling you to bookmark chapters, annotate text and use instant searches to find specific words or phrases. Several eBook files would fit on even a small laptop or PDA.

NEW: Save money by eSubscribing: cheap, online access to any eBook for as long as you need it.

Annual subscription packages

We now offer special low-cost bulk subscriptions to packages of eBooks in certain subject areas. These are available to libraries or to individuals.

For more information please contact webmaster.ebooks@tandf.co.uk

We're continually developing the eBook concept, so keep up to date by visiting the website.

www.eBookstore.tandf.co.uk